PRAISE FOR
EXTREMELY ONLINE

"Terrific. . . . [Taylor Lorenz] is a knowledgeable, opinionated guide to the ways internet fame has become fame, full stop. . . . *Extremely Online* aims to tell a sociological story, not a psychological one, and in its breadth it demonstrates a new cultural logic emerging out of 21st-century media chaos."
—Clay Shirky, *New York Times Book Review*

"Valuable and entertaining. . . . The book is an enlightening history of the pioneers of influencers."
—Andrew DeMillo, Associated Press

"Each story Lorenz spotlights is carefully chosen to highlight the power that users have historically held in shaping social media trends and culture."
—Pratika Katiyar, *Teen Vogue*

"If you want to understand what is happening on the internet, you start by reading Taylor Lorenz."
—Lucas Shaw, *Bloomberg*

"Tapping her deep expertise in the subject, Lorenz makes a strong case that creators—not the tech platforms—truly shaped internet culture."
—W. David Marx, *Washington Post*

"Fascinating, eye-opening, and unbelievable in equal parts, [*Extremely Online*] is a comprehensive witness account of the players, operatives, and principals driving content on the internet today, across platforms and even across entire industries. A must-read for anyone wishing to understand the most arcane conspiracy theories taking up the nation's consciousness to the sweeping cultural changes a rising generation has brought about."

—*Town & Country*

"More than just a history lesson, Lorenz's well-researched book does a better job of connecting the dots than almost anything else I've read on the subject of social media's meteoric growth, and the unexpected rise of the influencer."

—Lance Ulanoff, Tech Radar

"Readers will learn valuable lessons about . . . what goes viral and are sure to be blown away when they see the dollar amounts moving through the industry. This socioeconomics docudrama is both fun and terrifying, just like the internet."

—*Booklist*

"This astute debut from Lorenz, a *Washington Post* technology columnist, traces the tumultuous history of social media from the early 2000s to the present. . . . Lorenz accomplishes the difficult feat of wrangling a cogent narrative out of the unruliness of social media, while offering smart insight into how platforms affect their users. . . . It's a powerful assessment of how logging on has changed the world."

—*Publishers Weekly*

EXTREMELY ONLINE

THE UNTOLD STORY OF FAME, INFLUENCE, AND POWER ON THE INTERNET

TAYLOR LORENZ

SIMON & SCHUSTER PAPERBACKS
New York   Amsterdam/Antwerp   London   Toronto   Sydney   New Delhi

Simon & Schuster Paperbacks
An Imprint of Simon & Schuster, LLC
1230 Avenue of the Americas
New York, NY 10020

Copyright © 2023 by Taylor Lorenz

All rights reserved, including the right to reproduce this book or portions thereof in any form whatsoever. For information, address Simon & Schuster Subsidiary Rights Department, 1230 Avenue of the Americas, New York, NY 10020.

First Simon & Schuster trade paperback edition March 2025

SIMON & SCHUSTER PAPERBACKS and colophon are registered trademarks of Simon & Schuster, LLC

For information about special discounts for bulk purchases, please contact Simon & Schuster Special Sales at 1-866-506-1949 or business@simonandschuster.com.

The Simon & Schuster Speakers Bureau can bring authors to your live event. For more information or to book an event, contact the Simon & Schuster Speakers Bureau at 1-866-248-3049 or visit our website at www.simonspeakers.com.

Interior design by Lexy East

Manufactured in the United States of America

10 9 8 7 6 5 4 3 2 1

Library of Congress Control Number: 2023941468

ISBN 978-1-9821-4686-3
ISBN 978-1-9821-4687-0 (pbk)
ISBN 978-1-9821-4689-4 (ebook)

To Grandmom.

// CONTENTS //

Introduction: The Social Ranking ... 1

PART I
ONLINE INFLUENCE BEGINNINGS

Chapter 1   The Blogging Revolution ... 11
Chapter 2   The Mommy Bloggers ... 19
Chapter 3   The Friend Zone ... 29
Chapter 4   The New Celebrity ... 41

PART II
THE FIRST CREATORS

Chapter 5   The Rise of YouTube ... 59
Chapter 6   Creators Break Through ... 85

PART III
NEW DYNAMICS

Chapter 7   Twitter Follows Back ... 101
Chapter 8   Tumblr Famous ... 115
Chapter 9   Instagram's Influence ... 127

PART IV
THE PLATFORM BATTLES FOR CREATORS

Chapter 10	Vine Time	145
Chapter 11	A Tangle of Competitors, A New Era for Users	151
Chapter 12	Parallel Lines	161
Chapter 13	Counting Seconds	175
Chapter 14	The Shuffle	191

PART V
THE CREATOR BOOM

Chapter 15	The Winners	205
Chapter 16	Peak Instagram	215
Chapter 17	The Adpocalypse	229
Chapter 18	Breakdown and Burnout	245

PART VI
INFLUENCE EVERYWHERE

Chapter 19	TikTok Dominates	263
Chapter 20	Unlocked	273
Chapter 21	The Scramble and the Sprawl	283

Epilogue	289
Acknowledgments	293
Notes	297
Index	355

EXTREMELY
ONLINE

// INTRODUCTION //

The Social Ranking

THIS IS A BOOK ABOUT A REVOLUTION. LIKE MOST REVOLUTIONS, THIS one has done less than some of its vanguard promised and more than anyone predicted. It has radically upended how we've understood and interacted with our world. It has demolished traditional barriers and empowered millions who were previously marginalized. It has created vast new sectors of our economy while devastating legacy institutions. It is often dismissed by traditionalists as a vacant fad, when in fact it is the greatest and most disruptive change in modern capitalism.

When looking at what the internet has wrought in the last twenty years, we tend to focus on Big Tech: the massive corporations, the founders behind them, their visionary innovations, and the power they wield. But that's only half of the story. For all the platforms that Silicon Valley has created and algorithms they've tested, the real transformation has occurred closer to the ground. The business of Big Tech doesn't hinge on what they've invented but on what they've channeled. From the first amateur blog to the newest TikTok sensation, it has been users and those in their periphery who've brought the creative energy, the tech companies rising around them, fueled by the rich content and collective attention. It's users who revolutionized entirely new approaches to work, entertainment, fame, and ambition in the twenty-first century. And the first glimpse of this transformation came when the internet collided with the most status-conscious group: New York socialites.

※

New York socialites kept Who's Who lists as far back as the 1800s. Women like Gloria Vanderbilt, Nan Kempner, and Brooke Astor appeared in the pages of magazines, served as muses for fashion designers, and held prestigious jobs at publications like *Vogue*. At every party, a small group of trusted photographers and columnists was allowed entry, and media coverage was invariably aglow. While there had always been controversies and untoward behavior, it usually stayed within the circle.

But, on April 24, 2006, something popped this well-tended bubble. An interloper emerged. A mysterious website. A *blog*.

The site, socialrank.wordpress.com, came out of nowhere. On a website with a pale purple background, black text, and a banner of Champagne flutes appeared a list: "The Top 20 Female Socialites in Manhattan."

"It all started with a meeting in Four Seasons exactly two weeks ago," read the inaugural Socialite Rank post, "where a committee nailed down 132 finalists for the New York socialite power ranking system.... The group are the most praise-worthy, beautiful, press hogging, inspirational and hot personalities that make [the] New York party scene so ruthlessly competitive and yet so breathtakingly exciting."

The post explained that its Top 20 rankings were based on the evaluations made by judges in four categories:

- Personal style and designer relations (1–20 pts)
- Press coverage in major publications and gossip columns (10 pts)
- Appearances and commitment to events (10 pts)
- Hot factor—what makes each of the individuals sizzle with personality (10 pts)

Twenty-five women's names were listed, along with recent photos. At number one, a socialite by the name of Tinsley Mortimer.

Blogs dominated the internet at the time, but most were unfashionable. The web was still perceived to be for dweebs in sweatpants and thick glasses. Yet New York high society, while none would admit it, began to check Socialite Rank religiously. As word about the blog spread, some of the women in contention put more effort into their outfits; the demand for hot party invites grew. The rankings were built on points and metrics that, however subjective, could in theory be gamed. "All spots are up for grabs," the site declared. "You can sleep all you want after you die, but parties, pretty outfits and publicized engagements are what SR points are made of."

It was as much fun to speculate about the identity of the author(s) as it was to track the rankings. Socialites had always enjoyed playing out a fairy tale on the red carpet, but now they were cast as characters in someone else's play. *Whose* play was it, though? The anonymous posts relied on inside knowledge of parties, suggesting a mole. But the posts were also weirdly ungrammatical and voyeuristic.

Throughout the spring and summer of 2006, new socialite rankings dropped every other week, with near daily, gossipy blog posts in between. It was no surprise that on the list of top 20 Manhattan socialites, Tinsley Mortimer was a perennial front-runner. A Southern belle, she had debuted at the cotillion in Richmond, Virginia; attended Lawrenceville boarding school and Columbia University; and married into an oil-money fortune. She then embarked on a life of charity balls. While others in New York wore muted tones, she preferred pink, frilly outfits and baby-doll dresses. She also liked to wink at the camera in photos.

As the site's audience grew, the highs and lows intensified. The authors stirred up drama with every biweekly ranking. The comments section filled with insults, spats, and wild rumors of out-of-control behavior and drug abuse. Few outside New York had cared about high society before, but suddenly the whole scene was on display to anyone with an internet connection.

Within months, as *New York Magazine* reported, "Socialite Rank manipulated the city's gossip cycle, elevated unknown women to unlikely prominence, and gained thousands of readers, who filled the comment boards with catty and frequently venomous remarks."

Several women were devastated by the rankings and commentary. When the blog derided one socialite for "being a social operator and having a horse face," the insulted woman "cried for days," a friend recalled to *New York Magazine*. Yet in a hint of the dynamics that had already begun reshaping the media world beyond the Upper East Side, her tears later gave way to a welcome revelation. "Because of that site," the friend continued, "she's a huge social star." ("Social" here meaning within high society, since social media as we know it had not yet emerged).

By early 2007, Socialite Rank had been publishing its rankings for nearly a year. The holiday party circuit had come and gone, and the same women jostled for spots on the bi-weekly list. Mortimer remained the unchallenged queen.

Then, on February 8, 2007, the *New York Post* ran a story announcing an emerging new rivalry. The story declared that in a "scene straight out of 'Mean Girls,'" Tinsley Mortimer "looked less than delighted" at a recent fashion show when photographers asked her to embrace "up-and-comer" Olivia Palermo.

At first, Palermo received a warm reaction from Socialite Rank and started to climb the ladder. Like Mortimer, she had all the trappings of an A-lister: daughter of a real-estate developer and an interior decorator, she'd divided her childhood between New York City and Greenwich, Connecticut. Now she was living in the city studying fashion design at the New School.

Palermo was a constant presence in pictures posted by New York society's favored party photographer, Patrick McMullan. In rapid succession, she courted designers, signed on with charitable organizations, and hired a publicist—all the necessary socialite steps. But Socialite Rank compared Palermo to canned tuna, implying that she was packaged and inauthentic. The site pushed rumors and cruel gossip and tore her apart with misogynistic insults, calling her "desperate" and Mortimer "old," "washed up," and "trashy." It pitted the two women against each other in an imagined rivalry.

Within a year, the manufactured drama between the two women escalated, Socialite Rank sent the scene to a fever pitch. And then it disappeared.

Palermo's father was said to be seeking a court order to unmask the authors behind the blog. Before those threats came to fruition, *Avenue*'s Peter Davis had sleuthed out the answer. In April 2007, he confronted the masterminds behind Socialite Rank. It wasn't a group of top-tier socialites teaming up behind the scenes; in fact, it wasn't anyone most people knew at all. The all-powerful website was run by a completely random pair of Russian émigrés named Valentine Uhovski and Olga Rei. The duo was not remotely born into high society and they had essentially created the blog as a social experiment.

New York Magazine ran a cover story on the shocking reveal. Elite society was astonished. They realized they had been brought to their knees, to tears, to a frenzy—by two people they wouldn't have given the time of day. The outsiders had upended the ultimate insiders, and it had cost less than a trip to the hair salon.

The whole event was, to many, a strange blip. Socialite Rank existed for only a year. So why bother telling the story, almost two decades later? Because even after the site shut down, nothing returned to normal. Instead of New York society retreating back into its moated castle, the rupture only expanded. The story of Socialite Rank foreshadowed the next twenty years of online life.

The blog foreshadowed the rise of social media. Soon, we would all be beholden to public metrics, online rankings, pressures to commoditize ourselves and to build our brands online. The misogyny inherent to Socialite Rank and the way women were spoken about mirrors the type of sexism and biases women today still confront online.

Following the blog's implosion, Tinsley Mortimer and Olivia Palermo both did stints on reality TV, but Palermo emerged as the big winner. She used her role on MTV's *The City* to launch a career as an online fashion and lifestyle influencer. By 2016 she was reportedly being paid thirty thousand dollars just to walk a red carpet. Today, she has amassed over 8 million Instagram followers and juggles brand partnerships in between promoting her own beauty line, Olivia Palermo Beauty. She was the one, among all of those who had been pulled into the Socialite Rank craziness, who leveraged the new opportunities presented by the ever-evolving online world.

※

The internet has steadily changed everything around us: who we know; how we meet; how we work; how we date; how we play; who gets famous; whom we trust; what we want; and who we want to *be*. More and more of us receive the bulk of our information and entertainment through social media. Any of us, now, can spend years working to climb the career ladder, or we can aim to go viral and completely transform the trajectory of our lives overnight. The old world is gone. But it's not yet obvious where the new world will settle.

When the internet first emerged, Al Gore likened it to the highway system. After the highway network emerged, people needed new places to eat on the go, new places to stay cheaply. National chains appeared, but, in turn, mom-and-pop options vanished, and town centers dried up. Strip malls boomed and suburbs sprouted. The rhythm of neighborhoods, families, and daily lives changed—and those changes continued to play out for decades after the final mile of asphalt was set.

Thirty years later, Gore's pronouncement reads presciently optimistic and ominous. The internet brought its own tidal wave of developments. The first wave was the invention of the technology itself. Once the infrastructure was in place, once the computer sat on a desk in every home connected to the internet, the trillion-dollar question became: What would people *do* online?

In answer to that question, the YouTubes, Facebooks, Musical.lys, Twitches, and TikToks of the world emerged, each of them offering new ways to feel at home online, to see and be seen, to gossip and share the latest news. But their ascent took time. The world changed in fits and spurts, as we all figured out what we would do once we were online—and what, in turn, being extremely online would do to us.

We are still figuring that out. We are now decades into the digital age, but the ground still shifts daily. Who, with their dial-up connection, could have ever envisioned fake followers, TikTok news cycles, meme stocks, Instagram-induced plastic surgeries, or QAnon?

The old world of New York socialites collided with the future a few years before the rest of us did. Today's social media ecosystem looks like a larger, machine-run Socialite Rank. We're all now deeply cognizant of our status, our metrics, our potential for micro-fame or outright celebrity. Even if our goal isn't to be known to millions, we still fixate on the "likes" we get from friends or welcome new connections, even from people we hardly know. This phenomenon can inspire entrepreneurial fortunes, as well as nerve-racking anxiety.

This book is not a complete history of the internet and its effects on our lives—that would take tens of thousands of pages, if it could be written at all. Instead, *Extremely Online* offers a social history of social media. It is about a force and an industry that is upending legacy power, and the people—many far removed from Silicon Valley—who shaped this new landscape.

Nothing will ever be the same.

PART I

Online Influence Beginnings

// CHAPTER 1 //

The Blogging Revolution

Let's go back to the year 2000.

It was the year of the Y2K scare and the peak of the dot-com bubble. The web had been around since the early 1990s, long enough to spread euphoria and trepidation. The first browser had come out seven years earlier, but moving data over the internet was still arduous. Many ambitious companies hitched their cart to the proliferating internet, but their disruptive potential was qualified by many observers. Amazon, for example, was considered a threat to bookstores and music shops, but little else.

Millions of people were starting to enjoy being online. They logged on through portals like AOL, using dial-up modems that moved at a crawl. But once online, they could instant message and email friends, join chat rooms, and shop. They could even read articles from the few newspapers experimenting with putting their content online. Heavily pixelated photos, Flash animation, and ASCII art were as glitzy as it got at the time. (On a 56K modem, it would have taken twelve hours to download a single TikTok video). As for online clout, *Half-Life* forum admin was the best you could hope for.

That would all change, however, with the rise of the blog.

The "web log" originated in the '90s, when a cadre of early internet users began creating their own websites to share their thoughts and favorite links with the world. The barrier to entry was relatively high, since launching a website in those days required buying a domain name and knowing how to code.

That changed at the turn of the century, as blogging platforms like Blogger, Blogspot, and WordPress emerged. When it came to visual design, these platforms were unexciting. They offered cookie-cutter websites—usually text-only. But the bare-bones solution was sneakily revolutionary. Blogs could be set up in minutes. Suddenly, anyone with internet access could become a publisher. Media consumers became media producers.

It's hard to remember how novel this was. Before the blog era, if you wanted to share your ideas with the public, you had to make it past layers upon layers of legacy gatekeepers. Letters to the editor, call-ins to the radio, article or book submissions—all had to be approved by a faceless authority at a moated institution. Even for those who'd been admitted through the tall gates of legacy media, publication opportunities only presented themselves after years of rising through the ranks, flattering the powerful, and simply lucking out. You could always go it alone and create your own underground zine or DIY publications, but your reach was limited as long as the gatekeepers gate-kept.

Not so with blogs. You could say whatever you wanted, on any subject, in any style. For your entire life, you'd been an outsider. No longer.

Predictably, some of the first notable blogs focused on technology, and while their impact might have been large within the tech world, they rarely made an impression outside of it. However, in the political world, a blog's influence could extend beyond a narrow group of industry insiders, as shown by one blog with the somewhat cumbersome name *Talking Points Memo*.

Journalist Josh Marshall started *Talking Points Memo* days after the Bush-Gore election in 2000, when the result was still up in the air. He was covering politics for the bimonthly publication, the *American Prospect*. Marshall had some web-design chops, and he happened to have a vacation scheduled for the week after Election Day. As the Bush-Gore contest intensified, Marshall launched *Talking Points Memo* and posted commentary by the hour.

Marshall aggregated important news items and interspersed them with insider tips he received from fellow journalists and campaign officials. He seemed to post at lightspeed compared to everyone else, plus he could offer more color and candor than legacy media could. Soon Washington insiders were refreshing the site faster than Marshall could update it.

Talking Points Memo wasn't the first online site to cover politics by the minute. Years before Marshall launched his blog, a former CBS gift-shop manager by the name of Matt Drudge launched a political gossip newsletter called *Drudge Report*. While Drudge's website made a big splash—growing especially fast during the Clinton/Lewinsky scandal—Drudge was not a blogger but an aggregator. Marshall was playing a different game, reporting original stories and commenting in his unique voice. Within months, he decided to take the radical step of quitting his job at the *Prospect* to blog full-time.

Marshall got in at the right time, as he was among the first wave of bloggers who were attracting real audiences online. *Talking Points Memo* delved into the minutiae of policy debates and the Washington rumor mill. The writing was wonky and candid, not the view-from-nowhere type of writing that political news junkies were used to. That was the point.

As *Talking Points Memo* and its ilk got off the ground, the blogosphere bloomed around them. The total number of blogs doubled every six months. In 2006, there were 60 million blogs in existence. Blogging platforms expanded the web to non-techies, and soon new blogs emerged on everything—from indie music to Hollywood classics, to fashion, gaming, parenting, and drug culture—along with thousands of personal blogs that functioned as online journals.

Most legacy publications didn't see blogs as a threat at first. Bloggers looked like curious eccentrics, a band of second-rate scribblers with too much time on their hands. The old guard scoffed that bloggers' writing wasn't up to the standards of the *New York Times* or *Vanity Fair*. They doubted that bloggers could ever break consequential stories without the access and talent monopolized by legacy media.

Readers, on the other hand, enjoyed the lack of polish. The media environment of the 1990s was centralized and corporate after waves of mergers left only a handful of conglomerates whose content was middle-of-the-road, burnished, and safe. In 2002, *Wired* declared "The Blogging Revolution," a paradigm shift in how people distributed and received information: "Readers increasingly doubt the authority of the *Washington Post* or *National Review*, despite their grand-sounding titles and large staffs. They know that behind the curtain are fallible writers and editors who are no more inherently trustworthy than a lone blogger who has earned a reader's respect." Blogs offered readers everything that legacy media couldn't, revealing what writers really thought. What's more, blogs also enabled real-time interaction between writers and readers through comments sections attached to posts. Unlike message boards, blog posts primed the discussion with original, substantial content that was ripe for debate.

Soon little bubbles of taste, influence, and community formed, and they started to enter the mainstream.

It was from a reader tip that Marshall learned of a December 2002 toast given by Senate Majority Leader Trent Lott at the hundredth birthday tribute to longtime pro-segregationist Sen. Strom Thurmond. In his remarks, Lott openly praised Thurmond's overtly racist 1948 presidential campaign.

While the *Washington Post* and ABC News ran brief stories about Thurmond's birthday gathering, no outlet made particular note of Lott's remarks. After the reader's tip, however, Marshall ended up writing some twenty posts about Lott's speech and its aftermath. Marshall assembled a broad argument against Lott, citing similar remarks in Lott's past, establishing a pattern of praise for neo-Confederate causes and a refusal to condemn segregation. Soon, other bloggers, and then Washington media, took notice. Lott took to TV to try to repair his image, but his avoidance of apology only led to him being excoriated by both the left and the right.

Within two weeks of Marshall's first blog post, Lott resigned from his leadership position in disgrace. Washington insiders realized that the story would have never taken off if not for Marshall and the blogosphere. A blogger had just sacked the Senate majority leader.

On December 13, 2002, the *New York Post* ran the headline: "THE INTERNET'S FIRST SCALP."

*

Throughout the 2000s, in every field they touched, blogs circumvented gatekeepers and tore down old structures. Launching a blog required next to no monetary investment, which made it a venture within reach to all.

Here was the great advantage that promised to upend capitalism as it existed before the internet. With hard work and the cost of a few large pizzas, someone could take on a company with thousands of employees and hundreds of millions of dollars in annual revenue.

In 2002, a company called BlogAds launched to help blogs sell display ads on their sites. BlogAds was soon followed by Google's AdSense and other competitors, allowing top blogs to adopt an ad-driven business model similar to that used by print publications, but with vastly lower overhead and superior targeting capabilities.

Marshall started using BlogAds in 2003, and by the following year he was making nearly $10,000 per month. A few years later, ad revenue had grown so much that he was able to hire a team of reporters—often from old-media outlets—growing his staff to about twenty-five by 2012. A decade in, Marshall was no longer a reporter or a blogger; he was running a full-fledged media company. So it was elsewhere: popular sites like *Gawker* and *FiveThirtyEight* started with just one or two people, but as audiences grew, they were able to scale up into full newsrooms.

As bloggers proliferated, they didn't just adopt traditional beats; they shaped the cultures around the topics they wrote about. In 2005, Garrett Graff, who had helped run online outreach for Democratic presidential candidate Howard Dean, became the first blogger to receive a

White House press pass. That same year, Perez Hilton had Hollywood in knots, out-tabloiding the tabloids with his blog *PageSixSixSix*, which was later dubbed "Hollywood's Most-Hated Website." Music blogs like *Pitchfork* were defining an indie scene that the major labels scrambled to made sense of. Fashion blogs like the *Sartorialist* were identifying new looks before the glossy mags picked them up. Nightlife blogs like Hipster Runoff and photographer Mark Hunter's The Cobrasnake cultivated a new aesthetic for the era and launched internet "It" girls into mainstream celebrity culture.

In 2021, *W* magazine wrote, "Hunter's unfiltered nightlife shots defined an early digital aesthetic—and ushered in the social media age."

"My site was Instagram before there was Instagram," Hunter told me. It was the first time "regular" people were able to build their image off online party photos en masse. "There was this whole underbelly of nightlife getting documented and put out there," explained writer Lina Abascal, who documented the scene in her book *Never Be Alone Again*. "Sure, there were photographers at Studio 54, but there wasn't the internet . . . a whole night would be captured by people like Mark and uploaded online for people to go through." The model Cory Kennedy became a quintessential "Cobrasnake star."

By 2009, fashion bloggers like Bryan Boy and Garance Dore made their foray into high-brow fashion circles. Bloggers were suddenly sitting in coveted front-row seats during New York Fashion Week, then at Dolce & Gabbana's show at Milan Fashion Week, in a shocking upset that fashion insiders dubbed "blogger gate." "Bloggers have ascended from the nosebleed seats to the front row with such alacrity that a long-held social code among editors, one that prizes position and experience above outward displays of ambition or enjoyment, has practically been obliterated," wrote Eric Wilson of the *New York Times*.

As blogs boomed, traditional media felt the hurt, especially local and regional newspapers. Subscription rates everywhere plummeted now that the internet gave readers access to a wealth of free information, including articles from the very newspapers they no longer purchased in physical form. The industry's century-old business model crumbled, forcing newsrooms around the country to hemorrhage staff and shut down. As they did, gatekeepers went from dismissive to hostile. In testimony before Congress, David Simon, a former reporter for the *Baltimore Sun* and creator of *The Wire*, warned that the blogosphere was causing a media death spiral: "Readers acquire news from the aggregators and abandon its point of origin—namely the newspapers themselves. In short, the parasite is slowly killing the host."

By the end of the 2000s, it looked more like the parasite and host had merged. As top blogs expanded their headcount by hiring professional reporters, designers, and support staff, they came to resemble traditional media companies, complete with newsrooms and sales departments. Many legacy publishers realized that their best strategy was simply to invite bloggers in. Major publications, from the *New York Times* and the *Atlantic* to *Glamour* and *Elle*, hired the top crop of bloggers to fill out their ranks of writers and reporters. These same organizations also started major blogs of their own, or bought successful sites outright. By 2009, nearly half of the fifty most-trafficked blogs were owned by corporate media behemoths like CNN, ABC, and AOL. Yet while star bloggers in tech and politics received top billing, another class of bloggers was quietly ushering in a larger shift.

In the end, the defining figure of the blog era wasn't the nerd or the wonk. It was the mommy blogger.

// CHAPTER 2 //

The Mommy Bloggers

REBECCA WOOLF HAD HER FIRST CHILD AT AGE TWENTY-THREE, WITH a husband she hardly knew. It was 2005 and she'd moved from San Diego to Los Angeles six years earlier to work as a copywriter, ghostwriter, and headshot photographer. In her spare time, she blogged. She created her first blog in 2002 on Blogspot, an early publishing platform. Then she started a second one, "Pointy Toe Shoe Factory," where she wrote about traveling and single life.

When she got pregnant unexpectedly, Woolf wrote about her decision to keep the baby and marry the baby's father. She reflected on the drastic shift from carefree partying to impending parenthood. Shortly after their son, Archer, was born in May 2005, she decided to start a new blog cataloging motherhood. She named it "Childbearing Hipster" at first, then changed it to "Girl's Gone Child" a few months later.

Bloggers were writing on every subject under the sun in the early 2000s, and parenting was no exception. Blogs were offering content that couldn't be found in legacy media. As the *New York Times* declared, "The mommy bloggers were the first media voices who spoke directly—and exclusively—to mothers." With her new blog, Woolf joined a burgeoning class of mommy bloggers who had found their community online. The path that these young mothers blazed and the strategies that they developed would set the stage for online creators for decades to come.

The pioneer of mommy blogging was Heather Armstrong, a blond, attractive woman who started Dooce.com back in 2001. (The name "Dooce" came from her inability to spell "dude" during online chats with former coworkers.) Armstrong's blog soon developed a cult following.

Throughout the early 2000s, frank parenting talk was hard to come by. Women's magazines pushed an idealized, often misogynistic version of motherhood that was less and less relevant to modern mothers. Home life remained private, with family life and children considered personal matters. When mothers did discuss those things, it was often within the confines of each other's homes—not for the world to see.

Woolf and Armstrong were among the moms who, in grappling with changing but often contradictory gender roles, found themselves neglected by existing media channels. They had achieved successful careers before becoming mothers and held aspirations beyond homemaking. And now that they were stuck at home while their friends still lived the single life, they had complicated feelings about it.

So, a generation of mothers turned to the internet, either as readers, or writers, or both. Blogging gave them a needed outlet for their creative energy as well as a way to connect with others like them. What began as a hobby ultimately found millions of readers with a shared, unmet need: solace, entertainment, and camaraderie during a period of life that was often isolating and overwhelming.

Woolf, Armstrong, and their peers wrote deeply personal, raw, and unfiltered accounts of the sides of motherhood not found in parenting books. They spoke candidly about suffering from postpartum depression, struggling to breastfeed, and drinking wine during playtime. No topic was off limits. Many, led by Armstrong, wrote about their mental-health struggles.

"The early blogs were all about telling the messy story," explained Catherine Connors, who started the blog "Her Bad Mother" in 2005 and is now a writer and consultant in Los Angeles. "And there was a sense that, yes, it had to be really brutally honest if it was going to get an audience." Readers craved an antidote to the idealized depiction of motherhood that was championed by traditional media, and they found it in blogs.

Reading blog posts from that era today, it's shocking how mundane much of it was. Venting about a bad mealtime or stressful playdate was revelatory, but now, over a decade later, the chatty, swear-laden, unfiltered style seems completely normal—*because* mommy bloggers were the first to bring that honesty to the public sphere.

The emergence of mommy blogs was a form of liberation for women. Once they got online and found their people, writers like Woolf felt as if a cork had been popped. Finally, they could connect and express themselves without fear of judgment. "We all felt super empowered and fearless," Woolf told me. "We were all writing truthfully about our experiences as city moms, country moms. We created this virtual mommy group from all different backgrounds and shared this commonality of wanting to write truthfully about our lives and support each other."

As the blogosphere expanded throughout the mid-2000s, so did the "mamasphere." Mommy bloggers formed loose collaborative groups, cross-linking to other mothers and adding them to their "blog roll," a list of blogs linked as a list on one side of a website.

Blogs at the time were primarily text, so there were no performative photos or glossy presentations of these women's lives. Some readers didn't know what their favorite bloggers even looked like. That proved no obstacle to forming strong connections. The blogs were deeply personal in a way that almost no other media at the time could match. Devoted audiences and fandoms accrued as readers lived their favorite bloggers' adventures vicariously. "I had people show up at my home address: fans," Rebecca Woolf recalled.

By the end of the decade, Nielsen gauged that 20 percent of the online population was made up of "women between the ages of 25 and 54 with at least one child." Nielsen dubbed them "The Power Moms" and mapped out a core group of bloggers—"The Power Mom 50"— along with more than ten thousand smaller blogs that formed the larger network.

The next question was, "So what else can we do with this?"

Mommy bloggers realized that, in their numbers, they had untapped economic power. People wanted to read their work—a lot of people. That work took time and effort, sometimes as much as a full-time job, yet they were usually uncompensated.

BlogAds offered a profitable model for the biggest blogs—which by now included a few mommy blogs—but most bloggers didn't have enough traffic to bring in real revenue. Even for those who did, it was a fraught decision to try to make money online. Mommy bloggers' advantage was their authenticity. Ads were seen as tacky and inauthentic, a sign of selling out.

In 2004, Armstrong decided to start running ads on Dooce.com. She explained to her readers that generating income from the site would help with her family's financial pressures. "I've considered taking a job outside the home," she wrote, "but that would mean that I would probably have to give up this website. I don't possess the juggling skills to raise a baby and work a full or part-time job and maintain the amount of writing I have done here."

Despite Armstrong's trepidation and candor, her post received a tidal wave of backlash, with comments so cruel that she had to block them. "Fans were really pissed," she later told *Vox*. "They screamed, 'Who do you think you are?'" she said to the *New York Times*. "What made you important enough to make money on your Web site?"

Tech and politics blogs were already running ads, but when mothers started doing it, people became blind with rage. Armstrong was up against age-old stereotypes about women's work, that it was gauche for mothers to bring money into conversations about the labor they performed. Mommy bloggers were, first and foremost, mothers. Even though nearly every top mommy blogger worked on their blog full-time, they and their audiences appeared to internalize negative stereotypes about the economic value of the work they were doing.

Between seeking a paycheck and shutting down, Armstrong chose the former. Dooce.com became the first mommy blog to accept significant advertising. Ultimately, Armstrong found it "empowering," as she told *Vox* in 2019, "because I realized I didn't need some male executive in New York to tell me that my story's important enough to publish because I can just do it myself."

Running ads gave her license to treat blogging like a business and to be her own boss. She landed high-profile deals with Suave and other consumer brands. She partnered with a newly formed company called Federated Media, which sold ads and bundled blogs together. Soon the ad revenue made Armstrong the primary breadwinner of her household. Her husband quit his day job to help her run her blogging career.

※

Armstrong's decision to start running ads rippled across the mommy blogosphere. Other mothers began monetizing their blogs, too, and in doing so established a model for online creators that exists to this day.

The mommy blogosphere was an unexpected but perfect setting for a new path to online entrepreneurship. While the largest blogs on the internet aggregated links, mommy blogs were distinctly personality-driven and original, which favored a smaller scale in comparison. Many bloggers appreciated the built-in flexibility of blogging, since the work could be made to fit around parenting duties.

Meanwhile, consumer brands were interested in advertising to moms. Mothers are a multitrillion-dollar market in the United States alone, controlling as much as 80 percent of household spending. If you could get mothers talking about your product, you stood to make a windfall. In the early 2000s, Procter & Gamble launched a program where 600,000 mothers were given free samples of a product in exchange for mentioning it in everyday conversations with other mothers. In reviewing the results, P&G calculated that the average mother spoke to five people a day, while a "connector mom" spoke to twenty. By comparison, the right mommy blogger could reach hundreds or thousands of consumers with the click of a button.

This kind of breakthrough made marketers' eyes light up. Health and beauty brands started sending bloggers free products in exchange for coverage. (Many bloggers at the time never disclosed such exchanges, leading the Federal Trade Commission to release new disclosure guidelines.) A blogger didn't need enormous reach to make the cut. Nearly any online presence had greater reach than an analog connector mom.

While mommy bloggers shattered barriers, the demographic of women who were able to make it big in this new medium was skewed. The majority of those who were successful were attractive, thin, white women. Black mothers and mothers representing other marginalized identities were not granted the most lucrative brand deals. This bias went well beyond blogs and would become an even bigger concern as social media became more visual.

As their fame and revenue grew, mommy bloggers began hosting conferences with one another. One of the first and biggest events of this kind was BlogHer, founded in 2005 by entrepreneurs Elisa Camahort Page, Jory des Jardins, and Lisa Stone. There, and at similar events, women networked and traded information: tips on how best to manage your blog, create better content, engage with readers, and design one's site.

The BlogHer organizers realized that the camaraderie among attendees might translate into bargaining power. Women at the conference were already talking about how to set ad rates and demand monetary compensation rather than simply receiving free products for posts. BlogHer sought to formalize those bonds by launching an advertising network much like Federated Media. By 2007, BlogHer placed ads on over 1,000 women's blogs, helping more women chart a new path toward financial independence. One of the bloggers who took advantage of BlogHer was Ree Drummond, who'd launched a blog called the *Pioneer Woman* in 2006. Years later she became a multimillionaire tycoon overseeing a *Pioneer Woman* television show and spin-off cookbooks, and selling clothing and kitchenware.

✽

In 2008, knowing their audience was worth more than free detergent, bloggers made greater demands of their brand partners. Soon, Armstrong led the way with one of the first major "sponsored content" deals, getting Verizon to cover renovations to her office. Her husband shot a series of videos featuring Armstrong describing her new furniture and computer. Each one ran as a post on her site, opening and closing with the Verizon logo. To complete the look, the Armstrongs even had to switch from iPhones to Androids during the renovation.

For smaller bloggers, the perks *were* the treat. For example, *Forbes* reported in 2012, "Brand executives and women bloggers say the going rate for a $300 kitchen product is 500 monthly views; an all-expense trip to Hawaii requires at least 20,000 monthly views." But for the top class of mommy bloggers, the windfall brought in by ads and sponsored content was astonishing. "Overnight we all went from making practically $0 to $100,000," Woolf told me. "It happened so fast, but every one of us was like: Holy shit. We better strike while the iron is hot. Suddenly I had a video series with Target that I was getting paid a $20,000 retainer for." In 2011, the opportunities for monetization expanded further as new "affiliate marketing" programs allowed bloggers to earn commission on products they recommended.

With the arrival of Instagram, some mothers radically reinvented themselves, toning down their "real talk" in pursuit of a highly curated, visually pleasing feed. Some gritted their teeth and kept up their blogs. Others quit the internet altogether.

※

For Armstrong, it was already a difficult time. She went through a very public divorce in 2013 and struggled with mental health. She tried to push on with Dooce.com, even as the blogosphere emptied out and ad rates plummeted. But even she hit a breaking point later that year.

The issue came to a head during a 2013 Banana Republic campaign, when the company sent Armstrong and her assistant on a three-day getaway to Park City, Utah, complete with a horseback-riding expedition.

The company told Armstrong that they wanted her posts to be natural and fun. One of the things that had made Armstrong popular, however, was her propensity for a little TMI (too much information). After two hours in the saddle on the trip, she found the pain from horseback-riding so bad that she needed to practice breathing methods just to get through it. As she did, she couldn't stop picturing the natural-birthing book she'd read. In her post about the trip, Armstrong summed up the defining memory of her vacation with two words—"hairy vaginas."

Readers loved it. The story was laugh-out-loud funny. Banana Republic, on the other hand, hated it. If Armstrong didn't take it down, they told her agent at Federated Media, Banana Republic's parent company would pull its millions from Federated's network. Banana Republic eventually relented, but, from that point on, many companies insisted that sponsored content be approved by the brand in advance.

This new arrangement sucked the soul out of blogging for Armstrong. At the same time, her kids were reaching the age where they began to object to being the subject of her posts, sponsored or not. Armstrong was also grappling with the less desirable side of internet fame. Bloggers had started off looking for community, but once in the public eye, they became easy targets for what was already becoming a favored internet pastime: spewing hate, especially at women.

"This is where things started to get super dark and trolly," said Catherine Connors, another early mommy blogger. "Bloggers, especially female influencers, became targets of hate when they became personalities."

When mommy bloggers were covered in the press, legacy media reporters villainized them. The term "mommy blogger" itself was fraught and trivializing. Many women considered it demeaning and misogynistic. "It was used in the media pretty dismissively. The implication was always that these women are oversharing and profiting off their kids. People called them narcissists. It wasn't normalized and people were very uncomfortable with it," said Kathryn Jezer-Morton, who publishes a newsletter about the topic.

In order to survive this new landscape, a blogger had to be ready to face extreme vitriol daily. "A lot of bloggers who were super active when I was have left the industry," Catherine Connors reflected. "The spaces got more and more toxic and it seemed to be getting worse and worse." By 2015, for Heather Armstrong, there was only one solution: quit blogging. So, she did. A few years later she returned briefly, once again penning heartfelt and candid posts about her struggles with depression and alcohol addiction. Then, on May 9, 2023, after relapsing at her home in Salt Lake City, she took her own life. She died at age forty-seven.

※

"It's hard to overstate the way mommy bloggers changed motherhood discourse even into today," said Jezer-Morton. "Everything to do with boobs or feeding children, mommy bloggers really normalized that. A lot of these ideas of moms having fun in a way that's not totally squeaky-clean, mommy bloggers really pioneered that."

Ultimately, mommy bloggers, more than any other group, created the model that content creators and platforms cultivated in the decades that followed. They were among the first people to commodify themselves online, to post candidly about their personal lives—and then monetize. They had stumbled upon internet fame and blazed a path through it. Before long, others would follow, this time with more powerful tools.

// **CHAPTER 3** //

The Friend Zone

When we think of the internet today, we often think of tech behemoths like Twitter, YouTube, Instagram, and TikTok. It's practically gospel that social networks were designed to addict and captivate users. But the early years of social media were awkward and halting. Most platforms didn't know how to define what they were, let alone where they were headed. It's shocking to look back at how long it took for social media sites to embrace the power of celebrity. For its first decade, social media was stuck in the friend zone. Every social app in the late 2000s made sure you knew it was *a place to share with your friends!*—even as users were figuring out how to launch careers off of internet fame. Most platform companies refused to put any effort into cultivating creators until the mid-2010s, and by then it was only because creators had become big enough to force their hand.

Why, then, were the social media giants so late to realize what they had? Why not embrace the power of internet stardom from the start?

It all goes back to the first major contest for users between MySpace and Facebook. At that time, social media was a clever bit of tech in search of a purpose, and each company was pushing a very different vision. Their battle was winner-take-all: one would conquer, one would shutter. They were fighting over the future of entertainment and media, and their clash would reverberate for decades.

Digital social networks first emerged in the mid-'90s. In 1995, two Cornell students named Stephan Paternot and Todd Krizelman created TheGlobe.com. Their ambitious goal was to connect everybody on Earth through one website. They made headlines when the company went public on November 13, 1998, posting the largest first-day gain of any IPO to date. But the stock price plummeted a year later in the dot-com crash and TheGlobe.com never recovered.

In 1997, a site called SixDegrees.com allowed users to create a profile and connect with friends who were also on the site. As you added friends, the site would create a map of each user's network, allowing you to see your friends, your friends of friends, and so on. In its first two years, SixDegrees gained 3.5 million users before selling for $125 million at the height of the dot-com boom to YouthStream Media Networks, a marketing company.

These products were ahead of their time. Since they appeared before most people were even online, most users' friend maps didn't reflect their real-world social networks. The concept of making new friends online was still foreign to most people.

By the time Friendster launched in 2002, the internet was mainstream. The majority of Americans were online, with teenagers and college kids increasingly spending their free time there. Digital cameras were also widely available, which allowed users to easily post photos of one another. Users could create a profile and meet new people through mutual friends. Suddenly, people's amorphous web of relationships became visible, navigable. Some users found dates on Friendster; some reconnected with old friends and family. Others formed groups around niche hobbies and interests. Friendster became the hottest property in Silicon Valley since the dot-com crash in the late '90s, shifting investors' attention to social media.

But Friendster couldn't handle its rapid growth. The company didn't have enough servers in place, and just six months after its launch, it began to sputter under the traffic load. By the summer of 2003,

Friendster pages took twenty seconds to load, opening the door for competitors.

MySpace quickly stepped in. In August 2003, it started as a brazen clone of Friendster. MySpace's founders, Chris DeWolfe and Tom Anderson, had spent their tech careers immersed in the dregs of the industry, cobbling together copycat sites and making ends meet off of email spam networks and spyware schemes. The platform's only real innovations were that it let users rank their "Top 8" connections and create accounts under pseudonyms, whereas Friendster required users to register with real names. MySpace also cut a few features to make the site faster to load, and it accidentally introduced a fortuitous glitch in its code: Users could access the HTML code on their own pages, allowing them to customize their profiles with garish fonts, neon colors, animated headers, and background music. Young users loved it: they could make MySpace their own.

Friendster's woes gave MySpace a chance, and the upstart soon became the most popular social media site for teens in the United States. The founders of MySpace took every opportunity to accelerate growth. First, they made use of all the email lists that MySpace's parent company had access to. Next, they marketed the site at bars in Los Angeles, where MySpace was based. Anderson realized that MySpace offered a great source of free publicity for any rising musician or event venue, so he made a pitch wherever he went.

Finally, he courted a unique group of Friendster users. Some users of Friendster, like Tila Tequila, a Vietnamese pinup model, figured out a quick way to gain tens of thousands of friends on the site. Tequila would simply post revealing photos that violated the site's acceptable use policies. After the fifth time her Friendster account was shut down, MySpace's Tom Anderson made her an offer. He promised her that MySpace would support what she was doing, so she agreed to defect and invite her massive friend list over.

Tequila exploded on the platform, attracting a motley crew of teenagers, musicians, and models. The crowd made the site seem fresh and exciting, and by February 2004, after only six months in business, MySpace had accrued 1 million users.

One of the many young people who joined MySpace during its heyday was thirteen-year-old Kirsten "Kiki" Ostrenga. In 2006, in a quiet suburb in Coral Springs, Florida, she decided to set up an account. MySpace let you use a pseudonym, so she went by Kiki Kannibal.

At first, Ostrenga was simply looking for a respite from bullying. She was the new kid at her middle school, and she had a sense of style that didn't exactly fly under the radar. She chopped and teased her hair until it was twice its size, then dyed it hot pink. She wore dark eyeliner and heavy makeup with fishnet tights and lots of Hello Kitty gear. She was an easy target for bullies, and things got so bad that by eighth grade her parents withdrew her from school.

Homeschool afforded Ostrenga a break, but also boredom. With her parents' permission, she joined MySpace to connect with other kids her age. Soon she started posting photos of herself and her personal style. Ostrenga was thin, with pale skin and striking features. Her colorful outfits attracted attention on the internet.

She quickly found others who were embracing similar trends on MySpace. Their aesthetic was best described as "scene," a style that was decidedly different from mainstream culture. Ostrenga and teenage outsiders across the country could find their tribe on MySpace. Outcasts in their local schools, these users created a nationwide subculture.

At the time, MySpace profiles were public by default, but you could also request to connect as a friend. As she posted updates, Ostrenga began to get tons of friend requests. Every new connection seemed to bring a dozen more, and within three months she had more than 25,000 MySpace friends.

"It was kinda like a video game," she later told *Rolling Stone*. "I didn't see it as real people, more like as a number." As that number grew higher, Ostrenga found herself setting fashion trends. Along with a handful of other influential teen users, she was dubbed one of the site's "Scene Queens."

MySpace's founders had not anticipated the site's cultural importance. "In the span of just years, a generation of kids whose cultural consumption primarily came from TV and magazines were given a surplus of access to our peers, celebrities, and total strangers—including their likes and dislikes as they changed in real-time," observed *Input* magazine writer Sara Tardiff. "And for many, fashion subcultures were Myspace's most impactful export at the time."

While Ostrenga's friend list was ballooning, MySpace was rapidly establishing itself as social media's frontrunner. In two years the company went from 0 to more than 20 million users. It shot past Friendster, which ultimately crumpled under the weight of its server load. And in July 2005, News Corp bought MySpace for $580 million. The following year, News Corp signed an unprecedented ad deal with Google, promising MySpace and a few other properties over $900 million in revenue over the course of three years. Meanwhile, millions more users were signing on monthly. By 2007, MySpace was the most visited website in the world. By all objective measures, it looked to be the future of the internet.

※

In February 2004, Mark Zuckerberg launched TheFacebook.com at Harvard University. It was exactly what it sounded like—a digital version of the college's face book (a student directory with photos), albeit with some new features. Users could note their relationship status and the classes they were taking. They could send messages to one another and add a profile photo. In the tight-knit environment of a college campus, this was enough to upend social life at Harvard. By the end of spring, Facebook was a hit. And by the end of the summer, Zuckerberg had dropped out and devoted himself to spreading his social network from campus to campus.

Facebook remained limited to college students for its first two years. Users needed a .edu email address to create an account, and quickly students across many campuses joined. Unlike MySpace, Facebook's design was sparse and uniform. Students could create an account and populate their profile with basic information about their interests, classes, and relationship status. Then they could connect with their friends, post on one another's walls, and, beginning in 2005, upload photos and "tag" their friends.

Very quickly, Facebook became a staple of college life, making it easy to plan parties and see what all your friends (and crushes) were up to. In 2005 the site expanded, allowing high schoolers to sign up. That year, the user base grew from 1 million to 5.5 million. In 2006 it opened to the public, allowing anyone over the age of thirteen to join, ending the year with 12 million users.

Both MySpace and Facebook experienced exponential growth, but they also presented opposing visions for social media. MySpace looked like Times Square. It had gaudy, flashing profile pages, massive friend lists, pseudonymous accounts, lax moderation, and a salacious blend of musicians and models. Facebook, on the other hand, blended Ivy League exclusivity with Silicon Valley sleekness. Los Angeles–based MySpace was the social network where users could create identities anew; Harvard-originated Facebook asked for real names and educational affiliation. While MySpace pages were public by default, Facebook pages were walled off. MySpace had no friend limit; Facebook imposed a maximum of 5,000.

These differences were intentional, part of Facebook's core strategy. Zuckerberg sought to do what hadn't been done before on the internet: re-create *existing* social networks, online. He spoke of Facebook as a "utility," a breakthrough technology that would "connect the world."

MySpace's founders saw the value in online creators and talked about the platform as a media and entertainment company. "We want to be the MTV of the Internet," they told investors, seeing its top users as performers: "Unlike traditional media companies, MySpace generates free content, through its users; generates free traffic, by its users inviting their friends; and all you have to do is sell the ads!" Who could beat that?

Just as bloggers had stumbled upon the novel power of the internet to reach the masses, MySpace power users learned that they could use the scale of social media as a way to build and engage an audience. They "befriended" hundreds of thousands, and eventually millions, of fellow users. MySpace stars also understood the value of a direct relationship with their fans, as well as the importance of appearing relatable enough to allow fans to form parasocial bonds—a one-sided relationship where a fan develops a deep psychological connection to a media figure. As Tila Tequila told one reporter: "There's a million hot naked chicks on the Internet. There's a difference between those girls and me: Those chicks don't talk back to you."

While Tequila had come to MySpace with fame in mind, many of the Scene Queens—figures like Ostrenga, Hanna Beth Merjos, and Izzy Hilton—sought community and connection.

Many Scene Queens had brash personalities and opened up about highly personal details. Their MySpace friends followed their every move and became deeply invested in their lives. In their initially lonesome pursuit of connection, the Scene Queens amassed the most powerful emergent currency: online attention. For "a generation of suburban misfits," journalist Sandra Song wrote in 2019, they were "a crew of powerful online icons who helped morph internet culture into the beast it is today."

"We were basically influencers, but that term hadn't been created yet," Merjos said to *Input* magazine.

No road map existed for what these young female MySpace creators were doing. The musicians and comedians who'd gained prominence through the site could plug into an existing entertainment infrastructure of clubs, theaters, labels, and the like. This new breed of online creators broke the mold.

Despite their influence, most MySpace stars were unable to profit from their fame. Sponsored posts were starting to appear on blogs, but the popular personalities on MySpace were still considered too niche, young, or controversial for brands to work with. Some were recruited into modeling careers, while entrepreneurial users tried to leverage their reach to launch businesses or product lines. Such users hit a roadblock in the form of MySpace itself, however. The site forbade users from including any advertising that didn't come through the company's own channels, and the company's own channels wouldn't share any money with users.

A handful of MySpace personalities were able to navigate the gray area. Model Christine Dolce, for instance, with over 2.1 million friends, eventually launched her own fashion line and sold customizable dog tags, branding them with her pseudonym, "ForBiddeN." Others managed to parlay their online platform into mainstream opportunities like reality TV appearances. One of Hollywood's largest talent agencies, United Talent Agency, was the first of the big agencies to launch a digital division, which brokered a handful of early deals with MySpace stars. But most agents were more interested in pitching talent in the opposite direction, selling shows to upstart services like MySpace TV, an early web video service that eventually flopped.

As the biggest MySpace star, Tila Tequila had some of the most notable opportunities. Her MySpace fame eventually opened the door to a reality TV show on MTV, *A Shot at Love*. But the show was never a significant hit, and she drifted between diminishing D-list TV opportunities. Eventually, she resurfaced with another group of online proto-influencers—a troupe of alt-right neofascists who achieved notoriety in the run-up to the 2016 election. In her last major headline, Tila Tequila was photographed delivering a proud Nazi salute. She had done a lot to grab people's attention over the years—but this was the end of her time in the spotlight.

As for Kiki Ostrenga, she initially enjoyed the cultural relevance and power that came with her online persona. But the backlash was quick and harsh. In 2007, L.A. resident and twenty-eight-year-old pornographer Christopher Stone founded StickyDrama, a crowdsourced gossip blog that covered young scene kids who were gaining traction online. The site was particularly vicious and cruel to young teenage women. Ostrenga—sixteen years old at the time—was a frequent target.

StickyDrama attacked the Scene Queens while forums such as Get

Off My Internets (GOMI) and Blogsnark viciously targeted women bloggers whom they accused of acting like "attention whores." Anonymous users would torment and stalk young women they deemed "unworthy" of their followings.

This was the terrifying result of Ostrenga's early internet stardom. She became a character in the StickyDrama rumor mill. The online harassment spilled out into the real world. She was targeted by a sexual predator who serially sought out underage girls on MySpace. Ostrenga was doxxed and her home was vandalized, spray-painted with the word "slut." It took her years to dig out from the damage that becoming an "e-celebrity" had brought.

※

From the start, MySpace danced with a combustible combination of teenagers, models, bands, and a free-for-all atmosphere. The Wild West of the internet, it had been a refuge and a launchpad, but it also revealed the terrifying damage that the firehose of social media attention can inflict on a person. Facebook positioned itself as an alternative to the chaos. Zuckerberg offered the opportunity to connect without having to enter the tumult of the internet at large. By eschewing those seeking fame, Facebook grew into a juggernaut. It became a welcoming social hub that felt intimate, giving it an edge over MySpace.

By 2008, Facebook still had fewer users than MySpace, but it had momentum. MySpace was stumbling under the corporate umbrella of News Corp. Facebook moved fast and broke things, while MySpace struggled to add new features.

Facebook built tools that allowed developers to create businesses on its platform, including hugely popular games like Farmville or Words with Friends. Gaming was huge for Facebook. At one point, Farmville had almost 84 million monthly active users, with more than 34 million playing the app on any given day. Games like Farmville not only brought in new users but also normalized the passage of huge amounts of time on the internet.

Facebook moved its offices from Harvard to Silicon Valley, gaining access to world-class tech talent, while MySpace was still run by hustlers in L.A. Most importantly, Facebook made users feel comfortable online, while many felt increasingly alienated by MySpace.

By 2010, Facebook hadn't just dethroned MySpace, it had decimated it. A couple of years after becoming the biggest website in the world, MySpace was irrelevant. In 2011, it resold for $35 million, five years after its $580 million acquisition. The platform pivoted out of social networking completely to double down on music before fading into obscurity.

※

Throughout Silicon Valley, observers treated Facebook's win as proof of principle: Social media users wanted to hang out with their friends. Users who wanted something else—be they brands or fame-seekers—were a hazard that might eventually sink your platform.

Tech founders took this lesson to heart. In the late 2000s, the reigning wisdom was that social networks worked best when they were focused on pedestrian users and their friends, rather than wannabe stars and attention-seekers. You saw this reflected in dozens of start-up pitches. When Twitter launched in 2006, it marketed itself as a way for users to share their status with friends. When Instagram launched in 2010, it framed itself as a way for everyday users to share casual pictures with friends.

But in the decade ahead, the friend lesson would take some unlearning, as influential online creators forced Silicon Valley executives to recognize their value. There were plenty of reasons why Facebook rose and MySpace fell. But while MySpace's lost its allure as a platform, its early top users were on to something, something that it would take Zuckerberg a long time to realize.

Social media was practically a fame machine. Put millions of people in contact online, and certain personalities are going to stand out. They're going to ham it up, go viral, and attract a crowd. Content creators were the first to realize the cultural power latent in social media platforms.

Despite the lessons he ostensibly learned from the war against MySpace, Zuckerberg and his team unwittingly laid the foundation of influence status-seeking. In 2006, Facebook introduced News Feed, the first product designed to broadcast users' real-time updates and posts out to their connections. Prior to News Feed, users had to visit a friend's page to see their updates, which meant that no one would see a post unless they specifically sought it out. With News Feed, posts came to you in an infinite stream. This visibility led friends to share more, from status updates to weekend plans to breakup feelings. Every update had an audience, and even if that audience was just a handful of friends, someone was always watching.

The News Feed launch caused an uproar. Users complained that the platform was butting into their personal lives. "Stalkers Rejoice!" read a headline on Mashable. Massive Facebook groups formed in protest. In-person complaints even broke out at Facebook headquarters. But Zuckerberg stayed the course. News Feed boosted engagement, growth on the site, and virality. News Feed itself was ironically the very feature that allowed the uproar over News Feed to spread so quickly.

News Feed accelerated the internet toward the fame-based model that Facebook was originally opposed to. The feed architecture introduced millions of users to the engagement dynamics of the creator world. Everything you did online was now broadcast to a crowd and evaluated.

While Facebook was the product that introduced users to the broadcast nature of online content creation, its emphasis on real-world friendships rather than fan connections drove creators to build their audiences elsewhere. Facebook was the social media gateway drug for some, but seriously entrepreneurial creators found their biggest opportunities on the next generation of social media apps—YouTube, Twitter, and Instagram.

MySpace was irrelevant by then, though its founders' instincts were ultimately correct. Even as Silicon Valley pooh-poohed its model,

MySpace was wise to consider itself an entertainment company, exactly the way TikTok would position itself over a decade later. The MySpace founders saw the big breakthrough of social media, that anyone could now essentially run their own TV channel. Throughout MySpace's run, no one took its homegrown celebrities seriously; they were too far outside the mainstream, and too early. They were written off as freaks and narcissists, only to be ostracized and abused. Rival social media companies saw "e-celebs" as a liability rather than an asset. The nature of celebrity, however, was changing. The shift would not only transform the internet but the offline world too.

// CHAPTER 4 //

The New Celebrity

JULIA ALLISON WAS A JUNIOR AT GEORGETOWN UNIVERSITY IN 2002 when she started a dating column in the school paper called "Sex on the Hilltop." *Sex and the City* was one of the hottest shows on television. As her column became a campus sensation, Allison felt like Georgetown's own Carrie Bradshaw. The university's location in Washington, D.C., brought national coverage to her column. (When she wrote about dating an anonymous young congressman, the *Washington Post* was swift to reveal his identity.) Her peers enjoyed her candid style, but within months, her headlines began to enrage Georgetown alumni and some students. "I didn't write about sex very much," Allison told me, "but all the conservatives at Georgetown were so upset. I became this lightning rod." Still, Allison soon landed bylines at national outlets like *Cosmopolitan* and *Seventeen*. Film producer Aaron Spelling even optioned her life rights when she was twenty-one.

By the time she graduated and moved to New York City in 2004, Allison seemed poised for greater success. She had an undeniable magnetism, and she was unafraid to hustle. Her goal was to parlay her bylines to a writing career in New York City media. She even hoped to land her own TV show. That year, on a list of goals she brought to the city, she had written "become a cult figure."

Upon arriving in New York, Allison relentlessly emailed editors around the city. But she hit a brick wall. A few bylines and a college column weren't impressive enough to big legacy magazine editors in chief. Eventually, *AM New York*, a free daily paper, gave her a weekly column. The pay was $50 a week.

Allison got an idea when she saw Tom Wolfe on a book tour that year. Everywhere he went, he appeared in his iconic white suit. "He's a brand," she realized. "I've got to be known and become a name." Wolfe built his brand in another era, but he wasn't the only archetype for Allison to follow.

*

In the mid-2000s, Paris Hilton was often framed by the media as an anti–role model. Everyone knew her name, but they were told that you weren't supposed to want *that* kind of attention. "Few celebrities," wrote the *New York Times*, "have worked as hard at pure tabloid notoriety, or built reputations so unsullied by accomplishment or circumspection." But Hilton was playing her own game even then. Almost two decades later, she would look less like the know-nothing she played on *The Simple Life* and more like a visionary. Tabloids, reality TV, and the internet were all dissolving the barrier between the famous and the anonymous and, over the course of the next decade, Hilton used all three to vault herself onto the A-list. Sheeraz Hasan, the mogul behind Hollywood.tv, said, "All the things that people are doing today, with social media: The first person was Paris Hilton."

The great-granddaughter of Conrad Hilton, founder of Hilton Hotels, Paris Hilton grew up between New York City and Beverly Hills. Unlike her well-heeled peers who largely avoided media attention, Hilton ran toward it. She posed for photos at events across New York and Los Angeles. If there was a step and repeat, Paris was there. Her image became synonymous with the "McBling" style of the aughts: Juicy Couture tracksuits, rhinestones, Von Dutch trucker hats, and even "the accessory dog."

Older observers were hostile and bewildered. "A new brand of celebrity," Barbara Walters called her. "Famous for being famous." Walters herself was a "brand," the most recognizable and influential woman in the news business, a celebrity of such note to have been regularly imitated on *Saturday Night Live*. But Walters had worked her way up the ladder, fighting discrimination as a Jewish woman the whole way. She was famous because she'd *done* things, people said. And here, they complained, was this rich kid who had allegedly done *nothing*.

"Paris really started that movement of having paparazzi follow your every move," Kim Kardashian affirmed in the YouTube Originals documentary *This Is Paris*. "I wouldn't be here today if it hadn't been for her starting off in the reality world and her introducing me to that world. I think the best advice she ever could have given me was just watching her."

While Hilton courted tabloids, reality TV was remaking mainstream entertainment. Cable TV hit peak viewership in 2000, so there was huge demand to fill the hundreds of cable channels now on offer. Americans were watching less network news, making the economics of serious television journalism much more difficult. (Walters's frustration with Hilton was perhaps a frustration with the erosion of her own world.) Scripted dramas were expensive to develop, and younger audiences were tiring of stale, standard-issue laugh-track sitcoms. So, producers began turning to something new in the late '90s, with semiscripted reality shows like MTV's *The Real World*. For content-hungry executives, reality TV was cheap and simple.

Shows like *Survivor* and *Big Brother* were the breakout hits of the early 2000s, and networks rushed to acquire more reality content. Many early aughts reality stars were people with some sort of previous Hollywood legitimacy, like *The Osbournes* and *The Anna Nicole Show*, which both premiered in 2002. *Punk'd*, a practical-joke reality show, featured Ashton Kutcher, who at that point was widely known from his role on *That '70s Show*. *Newlyweds: Nick and Jessica* featured pop stars Jessica Simpson and Nick Lachey, premiering in 2003.

Eric Nies, one of the seven young adults who MTV cast on the first season of *Real World* in 1992, shot to fame because of the show. He was one of the first reality TV celebrities, and he proved popular enough that MTV tapped him to host a show called *The Grind*. But despite his rise, Nies was never able to cross from reality star to mainstream celebrity. He had largely disappeared from the tabloids by the late '90s, as had most others who appeared in similar series.

It took Paris Hilton to vault across that line—on the back of her own reality show, *The Simple Life*.

Hilton's tabloid escapades in the early 2000s caught the eye of Sharon Klein, then senior vice president of casting at Fox. Fox executives had been brainstorming reality-show concepts. One idea was a reality-show re-creation of the hit 1960s comedy *Green Acres*, about a rich New York couple forced to move to a farm. As the network hunted for the right Fifth-Ave-to-farm transplant, Klein invited Hilton in for a meeting. "I'm used to meeting with actors who are putting on a façade," Klein explained to Reality TV World. "She was so real. She was funny. . . . She was in her own reality and not embarrassed to talk about it."

When Klein pitched Hilton to her colleagues at Fox, it clicked. They would send Hilton to live and work on a farm, and they'd document every hilarious moment of it. Quite simply, said Klein, "they wanted to see stilettos in cow shit." They cast Nicole Richie as her sidekick, and on May 2, 2003, the network sent Hilton and Richie off to live with the Leding family in Altus, Arkansas, for a month.

When *The Simple Life* premiered on December 2, 2003, it was an instant hit. The show reached more than 13 million viewers, unheard-of numbers in reality TV at the time.

A couple of months before the show's release, Rick Salomon, an ex-boyfriend of Hilton's, over thirteen years older than her, leaked a sex tape of Hilton that he had recorded when she was nineteen years old, after allegedly coercing her into making it. At the time, Hilton said, she was recovering from physical and emotional abuse that occurred at a boarding school. Salomon sold the footage without Hilton's permission, right as *The Simple Life* was wrapping up filming. He reportedly made $10 million off the tape in the first year. Hilton, who had never given consent for it to be sold, made nothing. Devastated, she sued, which ended in a settlement.

The sex tape received national attention, from late-night punch lines to newspaper columns. The tape also broke the early internet. With no way to control its spread, the leak was a damage-control nightmare. In a previous era, Hilton might have lost her career due to the leak. In the early 2000s, however, the media controversy around the sex tape only added to her fame. Tabloids couldn't get enough of it, nor could burgeoning blogs like Perez Hilton's, which covered her relentlessly.

Paris Hilton was pilloried by millions. She became the butt of every joke. Yet *The Simple Life* didn't flop after the sex-tape scandal; it became a hit. Online, she became the most-searched individual in the United States. The discomfort of studio executives was irrelevant. Despite the vitriol, she developed a core fan base who stuck by her. She rode out the storm and continued to cater to the audience who understood her style and humor. They were out there, and she could increasingly connect with them directly—online.

Much of Hilton's rise came before social media was widespread, but she became one of the first figures to parlay her many headlines into a successful digital brand of her own. She used the tabloids, reality TV, and eventually the internet to ensure that her brand was ubiquitous. She began earning over $1 million per night for DJ gigs and raked in millions from fragrances and other merchandise. She created a persona that her fans admired—climbing the cultural ladder until she was a certified A-list celebrity.

Four years after *The Simple Life* premiered, Hilton's onetime assistant, Kim Kardashian, made her own debut as a reality star. *Keeping Up with the Kardashians* hit the air in 2007. The show gave Kardashian and her mother, Kris Jenner, the springboard to transform her family into an influencer factory. Like Hilton, the Kardashians seamlessly adapted to the internet and used it to skyrocket to billionaire status. Reality TV had primed them to share the details of their lives, to engage with fans, and to use every second of coverage to promote their own brands—personal and commercial.

To be sure, both Hilton and Kardashian began with real money and obscene levels of privilege. They required a small army of personal assistants, publicists, and media professionals to succeed. The fact that Hilton, Kardashian, and other heirs succeeded in reality TV wasn't shocking: the reality shows that worked best were filled with luxury lifestyles and dramatic characters, or with images of the comfortable spectacularly out of place. Reality shows were never intended to mimic the everyday reality of their viewers, and they didn't. But the world as presented through Paris Hilton and Kim Kardashian wasn't completely sealed off. In fact, it was increasingly accessible with just the click of a mouse or a visit to the right store.

The question, for anyone looking to learn from them, was this: Could you find the same success without the start-up capital and family name?

That's what the internet seemed to allow. That's what Julia Allison wanted to find out.

※

Allison started writing under her first and middle name, Allison, instead of her real last name, Baugher. She hustled, hard. She wrote her column for *AM New York*. She auditioned for and was even cast in reality-TV pilots. She appeared on TV to give dating advice on air. And in 2005, she started a blog.

The blog didn't come with any huge aspirations. Initially it began as a clearinghouse for what she couldn't put into her *AM New York* column. She wrote about things like her dating life and where she went to eat. Soon, she began posting pictures of her (quite affordable) outfits, something she dubbed "head to toes." Tumblr users would call these types of photos #gpoy, which stood for "gratuitous picture of yourself." Allison's posts were deeply relatable and spoke to Millennial women in ways that traditional women's magazines didn't. Like many bloggers, she found that the medium allowed her to develop a rapport with a small but growing audience.

Allison soon concluded that the print-media gigs she'd been chasing were a dead end. She leaned into her blog and by 2006, people started to pay attention.

At the time, *Gawker* was the most influential site for (and about) online media in New York. Allison flooded the blog's tip line with links to her own articles, and when she commented on *Gawker* stories, she would often include links to her own work.

Commenting relentlessly on someone's post to try to get attention for yourself is commonplace now, but it was galling back then. *Gawker* staff writers promptly chastised her for "gratuitous self-promotion."

Allison was undeterred. When she showed up to *Gawker* founder Nick Denton's 2006 Halloween party as a "condom fairy," in a dress she made out of prophylactic packages, he realized that she was no longer a figure the website could ignore. The next day, at Denton's request, *Gawker* writer Chris Mohney ran an 800-word article called "Field Guide: Julia Allison." The post was vicious, accusing her of attention-seeking in cruel language. (In true 2006 New York style, it even had a little barb thrown in about Allison being unknown to Patrick McMullan.) The subtext of the article was clear: Who does this woman think she is? Who decided *she* was influential? And yet, the article acknowledged, "she's everywhere, it seems."

The piece went viral. The hate-read attracted hateful comments. Allison was distraught, crying for three days and begging the editors to take it down. When they refused, she decided to respond. On her blog, she posted a photo of herself, derriere to the camera, sporting her condom-covered dress. "Dearest Gawker," she captioned it. "Kiss my ass."

It was the start of a long-running online rivalry between Allison and *Gawker* that brought more attention to both. One *Gawker* editor described Allison as "our Paris Hilton," a recurring figure whom everyone had an opinion about. To Allison's detractors, she was an undeserving "narcissist." To her fans, she was a savvy, self-deprecating woman trying to forge a new path for herself. To everyone, she was a new, enigmatic type of celebrity.

"There's a particular kind of fame that's very normal now," Allison told me. "But no one was prepared in that era. They used to call it micro fame, and it's this experience of blowing up when you don't expect to. You're not blowing up the way Taylor Swift blew up. Instead, you get a lot of attention and become a big celebrity—but only in a select niche. It creates a bizarre juxtaposition of being both super famous and unknown, all at once. The way people got well-known online, on the internet, was very new to everyone, including me."

✳

In February 2007, David Karp and Marco Arment founded the microblogging site Tumblr in New York. Tumblr allowed anyone with no coding experience to set up a beautiful, clean, and simple blog.

The idea for the site came from Karp's desire to start a blog that was sleek and visual. When he looked around, however, he couldn't find a platform to post "all these cool videos, links and projects that I wanted to put out there," he later told *.net* magazine. Part of his preference may have come from the fact that he was not a writer. The twenty-year-old high school dropout was a programmer and designer, so he decided to solve the problem himself along with Arment, a fellow developer.

They named their platform Tumblr. Within minutes, you could register any name you wanted @tumblr.com, pick one of many beautifully designed templates, and customize it. Then you could post text, images, GIFs, quotes, and videos.

Unlike its predecessors like WordPress and Blogger, Tumblr contained early social mechanics such as the ability to "reblog" posts. The reblog technology was developed by Michael Frumin and Jonah Peretti at Eyebeam Art and Technology Center in New York. Peretti's former roommate Tim Shey, whose company Next New Networks shared an office with Tumblr, compared reblogging to DJing. "You could mix in a little of your own voice here and there while still keeping the crowd happy," he wrote on his blog.

Tumblr soon became as much a social networking site as it was a blogging platform, allowing users to like, comment, and reshare content. Their timing was excellent, since MySpace had petered out; Twitter, founded less than a year earlier, was a niche, techy service; and Facebook, while dominant, still centered on offline friendships, not strangers with shared passions. Tumblr was messy, creative, full of pseudonymous users interacting with people they often didn't know IRL. It attracted creatives of all types and within two weeks of its launch, it gained 75,000 users. Julia Allison was among them and would post as often as ten times a day.

Before Los Angeles became the capital of the online creator industry, New York's "Silicon Alley" was where people building audiences on the internet congregated. By the late 2000s, more and more events cropped up seeking to manifest internet culture in "meatspace." In 2008, mayor Mike Bloomberg established "Internet Week" in New York City with the goal of "celebrating New York's internet industry." The city's growing Flatiron District was home to the nexus of Tumblr and blogger culture. Allison became a regular at Tumblr parties and a fixture of the NYC tech scene.

When Richard Blakeley, then head of video at *Gawker*, looked at the schedule for Internet Week in 2009, he saw a lot of panels but no after-parties. So, he decided to organize an event on the rooftop of the Empire Hotel on the Upper West Side.

With that, the "Webutante" Ball was born, an annual prom for the internet set. "Anybody who's anybody is at this party right this second," photographer Nick McGlynn, who cataloged the scene on his website RandomNightOut, told the *Daily Beast* in 2010. Allison, a "FameBall Hall of Fame" honoree, was on the "prom committee."

The Webutante Ball was hardly the only IRL internet event to appear. BuzzFeed, then an upstart digital media company, hosted meet-ups for its fans. In 2008, a group of Harvard students founded ROFLCon, a biennial convention dedicated to internet memes featuring a slew of early online creators dubbed "net celebs" by legacy media. There was also Social Media Week, founded in 2009, an event that hosted talks espousing the value of building an audience online.

That same year, internet culture news site Urlesque and Know Your Meme co-hosted a party called "A Night to ReMEMEber." The event, which attracted popular Tumblr creators, early YouTubers, bloggers, and internet enthusiasts, invited attendees to dress up as their favorite meme. The party proved to be so popular as memes emerged as a pervasive format on the internet, that Urlesque and Know Your Meme joined forces again months later to host the first "HallowMeme" party, this time encouraging attendees to dress as their favorite internet stars. The top three prizes for the best costumes that year went to a Three Wolf Moon group, a Keyboard & Cat couple, and a man dressed as a series of tubes (a dubious metaphor for the internet coined by a U.S. senator).

Allison was a regular at these sorts of events. She hobnobbed with other online creators, then referred to as "ceWEBrities." She documented the scene on Tumblr, and her posts racked up likes and reblogs. She used Tumblr the way that many people would eventually use Instagram. "I'd do my daily outfits," she said. "If I had an article out or a TV appearance, I would post about it. Or I'd post lots of behind-the-scenes things, for instance photos from backstage at Fashion Week, things people didn't normally get access to."

At the time, media observers didn't know how to react to what Allison was doing. Bloggers were a mainstream concept by the late aughts, but they had their own lane. Allison wasn't a familiar kind of author. She was a journalist who used photos, text, video—every medium available—to invite users into her world and build her brand. She talked about dating and sex in one breath and the trajectory of the tech world in the next. She called it "lifecasting."

Soon, you couldn't go anywhere on the internet without hearing about Julia Allison. The *New York Times* profiled the twenty-seven-year-old in March 2008. That July, she graced the cover of *Wired* magazine. Her columns were regularly teased on the cover of *Time Out New York*, and she made frequent TV appearances on all the major networks.

She was "famous for being famous," the media often declared, the same phrase they used for Paris Hilton. But producers kept calling, and editors saw that she got clicks—and sold magazines. Allison recognized this effect and saw an opportunity to flip her fame into a greater cultural and economic force.

"I figured out that when I posted things on my blog, I'd get emails about it the next day," she recalled. "For instance, I'd post about a swimsuit and get an email from the swimsuit company saying, 'Who are you? We're tracking all this traffic from your blog.'"

Getting a free swimsuit was trivial within the scope of her ambition. "I noticed Oprah had knighted these individuals like Suze Orman and Dr. Phil to be guiding forces in different areas," she observed. "I thought, That's not going to work with people my age, they're not watching *Oprah*, they're looking at the internet. So I thought maybe I could create people like that, but for the internet."

Allison raised funding from investors and launched a company called Non Society. "In retrospect, the name was awful, but we were trying to say we weren't normal society; we were rebels," she said. "I tried to find an apartment where we could all live that would be sponsored, where we could livestream or blog and shoot videos from." It was an early version of a collab house, a concept that would later explode in the 2010s. When Bravo commissioned a pilot for a reality show called *IT Girls* based on this concept, online commenters were up in arms. *She doesn't deserve it*, they argued, lobbing sexist attacks. *Page Six* called her an "internet fame-whore."

Nevertheless, Allison began pitching and landing bigger brand deals. In 2009, Cisco paid her $30,000 to create two videos at the Consumer Electronics Show. That same year she was paid $14,000 for four tweets about T-Mobile. Unilever's chief marketing officer, Simon Clift, asked Allison to speak to three hundred executives about influencer marketing, describing her work online as "a lesson for a $50-billion-plus behemoth like Unilever." Allison also signed a big deal with Sony to promote the company's new Vaio laptop and starred in Sony's ad campaign along with Peyton Manning and Justin Timberlake.

While her dedicated young female audience loved the campaign, *Gawker* trashed her in deeply misogynistic articles. One of them jeered, "Did you know Julia Allison carries a Sony Vaio Lifestyle PC in her purse? It's not the only thing she pulls out of her purse with the word Lifestyle on it," referring to a popular condom brand.

Nearly every article documenting Allison's rise contained a disturbing level of misogynistic language and tropes. Tech journalists, who were overwhelmingly men, implied that Allison was promiscuous. They used highly gendered language to slut-shame her and question her credibility as an expert on media and technology. She was accused of trying to sleep with the powerful men in tech whom she interviewed or partnered with. *Fast Company* ran a piece titled "Sometimes Breasts Aren't Enough, Julia Allison." *Wired* and the rest of the tech press was similarly hostile.

Like all misogyny, there was fear behind the fury. In a 2010 interview with *TheStreet*, Allison said, "I looked to the internet as a distribution channel . . . to allow me to cut out the middle men—the people who were running the magazines and who also took in the ad dollars." Media outlets had every reason to crush a pathbreaking outsider who didn't abide by wait-your-turn conventions or traditional hierarchies—especially a pathbreaking woman.

Despite the backlash, Allison amassed a fandom online that adored her. Young women in particular looked up to her and loved her upbeat personality and self-confidence. Brands, too, recognized her talent for building an audience.

As that audience grew, Allison spoke at major business conferences around the world. She attended the World Economic Forum meeting in Davos and the White House Correspondents' Dinner. She was the star of an event at the 92nd Street Y and gave a keynote talk at South by Southwest.

By the end of 2010, Allison moved to Los Angeles. She was cast in a Bravo reality show that seemed like a big break. It was called *Miss Advised*, about three single relationship experts attempting to balance their lives. But the show was a wash for Allison. She participated in one "painful" season and then vowed never to do reality TV again.

Looking back, it's astonishing that Allison kept going. People set up entire websites dedicated to smearing her. Some stalked her family members. Prominent journalists and commentators mocked her on national television and in the pages of major media outlets. *Radar* magazine named her the third-most-hated person on the internet, right above someone who tossed a puppy off a cliff in a YouTube video. When she met with investors, the media accused her of "canoodling," implying that she was having affairs with the men with whom she did business. Deals fell apart because of misogynistic, gossip-inspired blacklisting.

"The amount of hatred coming my way was distinctive and overwhelming," Allison told me. "No therapist was even trained in internet hate back then. It was hard to get help for the emotional breakdowns I was having from people lashing out at me."

By 2012, she decided that she couldn't handle any more online assaults. "It had been about ten years of my life, and I was exhausted," she said. "I felt beaten down, I felt completely disillusioned, and I wanted a different reality. More than anything, I wanted to be off the internet. I was like, 'I don't know how I'll make money, but I can't make it this way anymore.' And I never looked back."

She set out to erase herself from the internet. She spent hours deleting over 14,000 tweets, one by one. She removed Tumblr posts, made other accounts private, and restricted access to her viral YouTube and Vimeo videos.

Every so often she dipped her toe back in, and each time she regretted it. "I'd start to feel safe again and post a picture, then I'd get hate and delete it," she explained. And the hatred remained. ("If you follow JA now, her life at 40 is FULLY pathetic and unrealized. It is truly a valueless existence," wrote one user on Reddit in 2019.)

Recent years have brought a cultural reckoning with how young women of the '90s and '00s were treated by the media. Prominent women like Britney Spears, Monica Lewinsky, and Paris Hilton have appeared in docuseries that reexamine what they endured. Clips of misogynistic interviewers berating the women have been dredged up and posted to horrified young audiences on TikTok.

Allison, however, has not been blessed with such a reckoning. Instead, she has been ignored and passed over, unacknowledged for her pioneering role in the now $16.4 billion influencer industry. When Silicon Valley investors finally started paying attention to the online fame ecosystem, rebranding it the "creator economy," her name was never mentioned.

What Julia Allison did better than anyone in her generation was to leverage attention on the internet and shrewdly monetize it. These two practices are commonplace today, but in the mid-2000s, they were radical. Bloggers of the time cultivated a subject-matter niche to grow passionate audiences. But to try to break in with sheer force of personality, to use the new tools of the internet to go from no-name to mainstream success? Julia Allison was one of the first to even attempt it.

"[Allison] represented a moment when the culture of the web changed dramatically," the comedian Heather Gold tweeted. "She was the moment everything changed and that's where her significance to us lay. . . . She was not accepted at all by web culture. But what she did became commonplace."

"She used this medium and became unstoppable," former *Gawker* managing editor Choire Sicha said of Allison in *Wired*. "She just made it happen in a way that seemed seamless and kind of magical."

Paris Hilton changed the nature of A-list celebrity by becoming famous for being famous. Julia Allison changed the nature of celebrity by charting the same path with far fewer resources, with an internet connection as her main tool. Today, millions of users do exactly what Allison did. That she was able to forge the path while enduring relentless cruelty and abuse makes her achievement even more remarkable.

"People used to call me an attention whore, it's like, is that what you call authors who try to sell their book? Do you call a movie star that when they walk the red carpet? Would you ever call a man that?' I was trying to get people to read my columns so I could pay rent on my $2,500 studio apartment. Even that word, 'whore,' everyone uses it so much about me. She's an attention *whore*."

Allison lives a quieter life now. She resides in Cambridge, Massachusetts, with her fiancé and was recently accepted into a master's program at Harvard's Kennedy School in leadership and public policy. "I couldn't have a more different relationship with the internet now than I did ten years ago," she said, "and part of that makes me sad. I do think there's an inherent value to sharing vulnerably and authentically, and I get so much out of it when other people do that."

During the pandemic, she posted updates about her life on Instagram and Facebook—rarely, and only privately to a small group of select friends. "I do intend and pray to come back to the internet someday," she said, "even if it's as a grandma on TikTok. I will come back and be like, Fuck you all! But I'm still building up to that point."

PART II

The First Creators

// CHAPTER 5 //

The Rise of YouTube

IN EARLY 2005, A DATING WEBSITE TRANSFORMED THE INTERNET.

It started as a place where anyone could post short videos of themselves, where they could describe their personality, their ideal partner, their likes and dislikes. They could watch videos from potential matches and see if sparks flew. That was in theory. In reality, users were lucky if they got a single match. The founders became so desperate for users that they put out Craigslist ads, offering $20 to any woman who would upload dating videos to the site.

No one responded. Before it was out of infancy, YouTube looked like a failure.

Three PayPal alumni—Chad Hurley, Steve Chen, and Jawed Karim—founded YouTube. They poured money and time into web video technology, only to see their dating service flop. But soon they noticed people using their site in unexpected ways. Several users were storing home videos and camcorder footage on YouTube. Karim realized that, rather than expelling these users, they should lean into their activity. "We should just be a site where you can post videos of yourself," Karim emailed his co-founders. "Broadcast yourself. That's it."

The web had plenty of video-sharing options, including services from Google and Microsoft. But most were clunky. Users had to download proprietary video plug-ins, and even then, videos worked only on certain websites, which were often cluttered and confusing. You couldn't easily share videos across the web or embed them on your website.

YouTube's founders had inadvertently developed the perfect solution. By using Flash video, their videos worked well on any website. They'd also developed a user-friendly way to upload video files and share them easily.

By June, YouTube launched a beta version where any user could upload any video for free. The new slogan on its website read, "Your Digital Video Repository." It wasn't the most alluring catchphrase, but it did the job. The site picked up more users who "began using YouTube to share videos of all kinds. Their dogs, vacations, anything," recalled Jawed Karim.

YouTube gained 10,000 users within three months and was getting 100,000 views per day. Its initial growth was enough to earn it a $3.5 million jolt of early-stage funding from esteemed venture-capital firm Sequoia Capital. By the end of the year, YouTube abandoned dating completely.

On December 15, 2005, YouTube as we know it officially launched. Days later, it had its first massive hit. A comedy troupe called Lonely Island—made up of comedians Andy Samberg, Akiva Schaffer, and Jorma Taccone—had been struggling to get traction on their low-budget videos for years. The trio, who had met in junior high school, were working for *Saturday Night Live*: Schaffer and Taccone as writers, Samberg as a cast member. Playing to their strengths, they started experimenting with online video, and their latest *SNL* bit was a "digital short," a pre-recorded clip that aired in between the show's live skits.

In the video, "Lazy Sunday," Samberg and fellow cast member Chris Parnell travel around New York City rapping aggressively about a leisurely weekend afternoon that included getting cupcakes from Magnolia Bakery (famous from *Sex and the City*), going to a matinee screening of *The Chronicles of Narnia*, and a trip to a bodega. They shot it themselves using a camcorder.

Within hours of "Lazy Sunday" airing on *SNL*, a bootleg copy of it was uploaded to YouTube. YouTube was still relatively unknown ("We'd never heard of YouTube until 'Lazy Sunday' came out," one member of Lonely Island told *Variety*), but within a week, "Lazy Sunday" amassed over 2 million views, becoming the first TV programming to find a second life online. YouTube's total traffic went up 83 percent that week. The clip made the site a phenomenon.

"Lazy Sunday" became a blueprint for viral success and signaled a shift in online video. Digital video cameras had only recently become widely affordable. Suddenly, anyone could grab one and shoot a short film that could be seen by millions. Thousands of copycat videos followed, along with myriad other types of videos set to music—parody songs, skits, lip dubs. Anything a bored teenager could dream up on a lazy Sunday afternoon could now be broadcast online.

After "Lazy Sunday," YouTube's growth skyrocketed. In February 2006, the site reached 25 million views per day, up from 3 million in December. NBC Universal soon expressed its discontent, however, launching a raft of copyright-infringement lawsuits against YouTube. They ordered all illegal uploads of the digital short stripped from the platform. Hollywood had watched the internet destroy the music industry through file-sharing sites like Napster, and it wasn't going to suffer the same fate without a fight. Thus YouTube's first hit was also its first major copyright woe, one of the many it would face over the coming years.

Nevertheless, the platform viewed the episode as a win. The viral *SNL* clip established YouTube as the dominant player in online video, and its competitors were caught on their heels.

※

From afar, George Strompolos, a partnerships manager at Google, was watching YouTube's ascent in dismay. He had joined Google in 2004 at age twenty-four, after working at *Wired* and CNET. He was given the role of partnerships manager for Google Video, which was one of several services under the purview of executive Susan Wojcicki.

The company's initial goal with Google Video was to create a search engine for video, just as they had for the web. Its strategy was to court major corporate clients to hash out licensing deals with studios and labels, to make Google an online hub for TV and movies.

These deals were complicated, however, and studios weren't leaping at the chance to give up rights to their video catalog. As Google Video focused on rights acquisition, they weren't paying as much attention to its everyday users. Google Video didn't have the social features that were built into YouTube, such as top user videos on the home page, allowing for discovery and connection. YouTube's comments and subscriptions features made it the best place for users to share original content and find a community.

People flocked to the platform. They started treating their YouTube pages as channels—not just simple storage for video but a steady stream of video updates, where viewers could tune in and interact with the creators. YouTube created a whole new class of "video bloggers," or vloggers, who turned on their webcams and just started talking.

Across the country, people opened up about their daily lives into their webcams, making heartfelt confessionals, ranting, and goofing off. These video diaries captured the same freewheeling energy as the blog boom, but with a new level of immediacy and access.

The mainstream media dismissed vlogging in 2006, if they mentioned it at all. Vlogs were amateurish, with grainy video and lots of dead air. Most videos lacked narrative, drama, or any production value whatsoever. Media observers had yet to recognize the power of blogs or the internet, and if they did take notice of online video, it was to note it as an oddity—something only a niche group of weirdos partook in.

That all changed thanks to a self-described dorky sixteen-year-old YouTube vlogger named Bree Avery, who posted under the handle Lonelygirl15. Bree posted her first video in June 2006, sitting at the edge of her bed on top of a pink bedspread with a green teddy bear in view. She didn't want to reveal her exact location, but said that her hometown was boring, "Like, really, really boring. That's probably why I spend so much time on my computer."

Bree's videos were strangely captivating. She began posting regularly, usually from her bedroom, with her floppy-haired friend Daniel sprawled in the background on her bed. She spoke candidly about the highs and lows of teenage life: the stress of homework, spats with her parents, her relationship with Daniel. Bree was charming and funny, and people formed a bond with her as they literally peered into her bedroom. Soon there was drama between Bree and Daniel, along with intrigue about her parents, who were supposed to be involved in a mysterious religious sect.

Lonelygirl15 was the talk of the internet in 2006, and soon became the scandal of the year. The channel, it turned out, was an elaborate work of fiction.

Lonelygirl15 was the brainchild of an unlikely group: Mesh Flinders, a twenty-seven-year-old screenwriter; a former doctor named Miles Beckett; and husband-and-wife lawyers Greg and Amanda Goodfried.

Beckett discovered YouTube, like many others, after watching "Lazy Sunday." He was inspired to create something for the platform. It wasn't until he met Flinders at a birthday party in a Los Angeles bar that a project took shape. For years, Flinders had been nursing an idea for a story about a friendless homeschooled girl whose parents had fallen victim to a strange cult. He'd written a screenplay but wasn't getting any takers. When Flinders shared the idea with Beckett, they decided to collaborate. They cooked up a script and within two weeks had an entire content calendar laid out for three months' worth of Lonelygirl15 videos.

To handle the business side of the project, Flinders and Beckett brought in Goodfried, a lawyer who became the show's producer. His wife, Amanda, also joined the project and ran "Bree's" MySpace page.

"If we didn't do it, then someone else would," said Beckett. "Somebody was going to create a scripted show on YouTube that uses the vlogger format, and if they were marketing savvy they would make it feel real so there would be talk about it."

After posting an ad on Craigslist for an actor, the team landed on Jessica Lee Rose, an actor and recent graduate from New York Film Academy in Burbank. Rose had a MySpace profile at the time, but no internet presence beyond that. She hadn't done any television or commercial work before. She was nineteen but could pass for younger. She had even been partly homeschooled.

Rose was wary at first. "When you're 18, 19, you think about going to L.A., and how you're going to be in a movie or a TV show," she later told the *Guardian*. "Then I found out it wasn't a movie and it was going to be on the internet. I instantly thought, this is some scam. I really thought this was what I'd been warned about moving to Los Angeles." They were able to convince her to sign on, however, and she deleted her public MySpace profile before shooting began.

Her castmate who played Daniel, Yousef Abu-Taleb, was a bartender and waiter. He also responded to a Craigslist ad. His average looks won him the part, as no one involved with the project believed that a more conventionally handsome young man would spend his time posting vlogs to YouTube.

Flinders and Beckett had consumed hundreds of hours of YouTube and studied what made videos pop. They discovered the power of the comments section. For Bree's videos, they instigated discussion and replied to everyone. Their efforts landed Lonelygirl15 a spot on YouTube's "most commented" section of the homepage, winning them views and subscribers.

They also figured out the precise frame in a video that YouTube would pick as the thumbnail. "If it was a good freeze frame, you would get like 100,000 more views," Flinders noted.

Lonelygirl15 soon dominated YouTube's "most viewed" section, with the videos amassing up to half a million views. After just a couple of months, Lonelygirl15 became the most-subscribed channel on YouTube. By the end of summer, the show was too big to escape scrutiny. A lonelygirl15 forum appeared, where fans could speculate about her background. Meanwhile, the project started to impinge on Rose's offline life. She was recognized in public once, but nothing came of it (camera phones and social media accounts were not yet universal). Then someone noticed that the fan site itself, lonelygirl15.com, had been registered before Bree began posting—and it was registered to Beckett and Flinders.

People began openly wondering if the whole thing was an advertising stunt or elaborate hoax. Many thought it could be a promotion for a horror film. One fan of the channel embedded an IP tracker on a fake MySpace profile and messaged Bree. When Amanda Goodfried responded from the office of her day job at Creative Artists Agency, the online sleuths leaked the info to *L.A. Times* journalist Richard Rushfield, who ran a story revealing the link between Lonelygirl15 to CAA. Shortly after that, an eighteen-year-old named Matthew Foremski found a cached version of Rose's old MySpace profile. His father, Tom Foremski, was a journalist who ran the blog *Silicon Valley Watcher* and published a piece with his son. They had uncovered the identity of Lonelygirl15.

The *L.A. Times* soon revealed the team behind Lonelygirl15. "It turns out the people behind the wildly popular website lonelygirl15 are not studio executives, Internet moguls or, as some suspected, Satanists," declared the September 2006 story. "Instead, they are aspiring filmmakers who met at a mutual friend's birthday party in April."

Rose braced for a wave of backlash. "The night they told me they've found your name, they know who you are, they've found your pictures on the internet, I thought: Oh my God, are people going to hate me? How is this going to be perceived?"

Instead, the big reveal only generated more excitement. People loved it. The creators were signed by CAA, marking a major foray into online talent for the agency. The series ran for another two years. Abu-Taleb continued to play Daniel, and Rose reprised her role as Bree until she was killed off in the second season. Rose even graced the cover of *Wired*, which ran a major feature on the show.

What stood out to Flinders was how simple it was to put the project together. "We did this with zero resources. Anybody could do what we did," he told the *L.A. Times*. The sum of their equipment was two desk lamps (one of which was broken), an open window, and a $130 camera.

YouTube had yet to roll out monetization during the first season, so the group wasn't initially making money off the views they raked in. But in later seasons, the show became profitable, blazing a path for future web series to follow. The creators formed a production company that sold product-placement ads for the series. In 2007, it was one of the first YouTube series to feature a sponsorship (from Neutrogena). The deal was a milestone in the online creator world, showing that youth-focused brands were beginning to take notice of the emerging platform.

※

YouTube's growth accelerated as Lonelygirl15 made headlines. Over the summer of 2006, YouTube became the fastest-growing website in the world. And while Lonelygirl15 remained the top channel on the site, YouTubers of all ages were attracting viewers, from teenagers to octogenarians. (Just before Bree took the top spot, the most-subscribed channel belonged to Peter Oakley, aka geriatric1927, a British retiree who posted videos discussing his life as a widower and recounting memories of his time as a radar mechanic during World War II.) The people posting on YouTube were the type of people whom Hollywood and mainstream media typically ignored.

YouTube's founders couldn't have envisioned how quickly the site would grow and diversify. As they saw audiences form communities, they rolled out social features to help users engage with one another and stay up to date on their favorite YouTubers. Editorial staff also fielded monthly submissions for the homepage's featured videos section, adding a curation layer that helped up-and-coming creators grow and go viral. YouTube was still far from cool, but it was happening.

By mid-2006, Susan Wojcicki, the executive who oversaw Google Video for years, had a tough choice to make. She either needed to revamp her product from the ground up to compete with YouTube or admit defeat and cut her losses. After weeks of deliberation, she chose the second option—with a twist. Wojcicki was convinced that online video was revolutionary. Rather than throwing an astronomical sum into saving Google Video, she decided to throw it at YouTube instead.

Wojcicki put together a pitch to buy YouTube outright. She and Google leadership went back and forth, ultimately landing on an offer the start-up couldn't refuse. In October 2006, Google purchased YouTube for $1.65 billion.

At the time, many observers derided the deal as folly. Its focus on user-generated content had brought incredible user growth, but it was being sued for copyright infringement by practically every network, studio, and label in show business. If any of those lawsuits succeeded, it could doom the company.

Google was taking a big bet. But when it looked at YouTube's rise, it saw pure potential.

Perhaps no one recognized its promise as clearly as Google Video's partnerships manager, George Strompolos. The day after YouTube's acquisition, Google CEO Eric Schmidt asked Strompolos and several of his colleagues to join the YouTube team. Strompolos was thrilled.

"There was this idea that you could create your own channel or station, and you didn't need the traditional media to put you on TV anymore," he recalled. "YouTube was faster and looser with what you could upload, and it was growing really fast." YouTube was still very much a start-up at the time of the acquisition. The whole company consisted of barely more than fifty people and operated out of offices above a pizza shop. The vast majority of the staff were engineers, which made Strompolos, with his background in media, feel like an intruder. When he entered YouTube's office, though, he brought a new perspective, one that saw YouTube's potential not in its storage capacity or buffering algorithms but in its creator community.

Chad Hurley, YouTube's co-founder and CEO, was also thinking big about YouTube's broad user base. When Hurley announced that YouTube was considering sharing ad revenues with users, Strompolos leapt at the idea. People were already making incredible art on the platform with no resources; imagine what they could create when they had some. For the rest of 2006 and into 2007, they built what would eventually become one of the first ways for social media creators to earn money from their work: the YouTube Partner Program.

※

While Strompolos and his team got to work, the YouTube platform was regularly plucking creators out of obscurity and granting them audiences of millions. Memes existed from the birth of the internet, but YouTube introduced a new sense of pace and possibility to internet fame, becoming a flywheel for viral content. Top videos launched creators to a level of online fame that hadn't seemed possible only years before. Once they acquired that fame, however, there was no real playbook for success. No one knew how to predict web fame, let alone what to do when you achieved it.

In 2006, Adam Bahner was a twenty-four-year-old self-described nerd in his third year of a PhD program in American Studies at the University of Minnesota when he discovered YouTube. He'd pursued music as a hobby, hauling his keyboard through deep Minneapolis snow to play at open mic nights. The crowds were never more than a handful of people, but he loved it. When he saw internet video arrive, he thought that putting his music online might get him more exposure. At least he wouldn't have to trek through a blizzard.

Bahner constructed a makeshift studio in his living room. He built a large wooden frame and hung bed sheets from it to form a backdrop. He ordered his first new computer in seven years to handle music and video editing. In January 2007, he began uploading content under the stage name Tay Zonday.

In his first video, he sang the gospel song "Swing Low, Sweet Chariot" in a low register. "I got the honest feedback you get on the internet in the comments," Bahner said. "People were like, 'My ears are bleeding.' Half was negative and half was positive."

Even the modest views on his first video amounted to a bigger audience than the open mic nights, so he continued, gaining viewers with every video. In April, Michele Flannery, then YouTube's head of music, emailed Bahner. She'd come across his videos and wanted to let him know that a song he'd uploaded called "Love" would be featured on YouTube's front page.

"Being featured on YouTube's front page in 2007—that was like a Grammy or an Emmy," said Bahner. "That high level of exposure for a viral video back then—it was like winning the lottery as a content creator."

Preparing to take full advantage of the coming attention, Bahner rushed to finish a ballad he'd been working on. He'd written the lyrics over the course of six weeks, linking them to a piano riff that had been running through his head. On April 22, 2007, he uploaded the song "Chocolate Rain."

"Chocolate Rain" was a ballad about systemic racism and Bahner's experience as a young Black man. "Every line of the song can be broken down into this critique of how we continue to avoid confronting institutional racism in this country," Bahner said. At the time, almost no one noticed what exactly he was singing about but, in increasing numbers, they noticed him singing.

"Chocolate Rain" earned a quick 30,000 views, a respectable number for the time, and he received another wave of attention from the YouTube home page. But the real traffic came a couple months later. In June 2007, a follower of his uploaded "Chocolate Rain" onto Digg.com, a popular aggregation site. "Chocolate Rain" hit the top of Digg.com and stayed there for two straight weeks, bringing hundreds of thousands of views to Bahner's performance.

After it appeared on Digg, someone shared it on the platform 4chan, which in turn kicked off a crusade to make "Chocolate Rain" a mega hit. The collective users on 4chan's /b/ message board (the board for random discussion) were incredibly adept at manipulating the early internet. They'd game systems to make content go viral and shine a spotlight on internet oddities that they believed deserved a larger audience. Something about Bahner resonated with them. He was autistic and in many ways the opposite of an artist who'd compete on *American Idol*. But there he was, holding nothing back. With his square glasses and earnestness, he became a hero to the type of (mostly) men who populated online message boards in 2007.

4chan users pulled pranks like calling into a talk show and singing "Chocolate Rain" loudly over the phone. Engagement on YouTube was not always positive. Some viewers spammed Bahner's video with racist epithets and taunted him in the comments. But nothing stopped "Chocolate Rain" from spreading. It became an anthem for power users of the internet, shooting Bahner to overnight fame.

YouTube embraced "Chocolate Rain." The company helped Bahner orchestrate a press tour, and he appeared on cable news and late-night shows. Thousands of people created remixed versions or recorded dances to the song in their bedrooms. That August, YouTube did an entire homepage takeover where every video was a parody of the song (Bahner's original video being the one prominent exception). Previously, MySpace had been the undisputed internet hub for music, but the multifaceted promotion of "Chocolate Rain" left no doubt that the next big song would debut on YouTube.

Even with all the attention, Bahner was hardly able to capitalize on his success. YouTube's monetization efforts were still nascent, and brands were still unsure of how, or whether, to leverage viral celebrity. Bahner had no professional background in entertainment, and he failed to treat the song itself as an asset.

"I made terrible business decisions at that time," he admitted. He wasn't thinking of revenue opportunities, because no one who had gone viral had ever really made much money before. For instance, rather than make "Chocolate Rain" available for purchase on iTunes, he uploaded the song for free to an MP3 sharing website.

He got thousands of incoming requests to appear at everything from birthday parties to corporate gatherings. Three major labels reached out wanting to sign him. Yet it was still hard to decipher which opportunities were lasting and real, and which were diversions.

Bahner ended up declining the majority of requests, but he did say yes to a Dr Pepper ad, one of the more obviously legitimate and exciting opportunities. It wasn't much money, mostly just an adventure he could have never imagined a year earlier. But overall, the whole experience was a lesson in how difficult it was to harness web fame and channel it into something of lasting value.

Bahner had a choice to make: Was it time to move on from internet fame? He decided it was not. In 2008, he left his PhD program with a master's degree. He would commit himself to building his YouTube career full time. The only question was: *How?*

※

True to the spirit of early YouTube, the most iconic clip of this era was a cat video. In the 1980s, Charlie Schmidt, a multimedia artist, was bored, with an old Beta camcorder and his cat Fatso. He made a goofy video where it looked like his cat was playing a song on his electric keyboard. Decades later, in June 2007, Schmidt uploaded that video to YouTube under the title "Cool Cat," and initially it failed to garner much traffic. "It started getting like 30 hits a day," Charlie told Mashable. "To me, 30 hits a year would have been a big deal. So I was happy. I felt rich and famous."

In 2009 a man named Brad O'Farrell from the multi-channel network My Damn Channel (now Omnivision Entertainment) reached out to him. O'Farrell thought that Schmidt and Keyboard Cat had the potential to go viral just like Adam Bahner had, if given a little boost. He asked permission from Schmidt to create a mashup of the video then sent it around to some friends in the NYC internet scene. Almost overnight, the video of Fatso playing the keyboard blew up.

"Within three days, my YouTube channel went absolutely bonkers. It was doing 10,000, 20,000, 30,000 hits a day," Schmidt recalled to Mashable. "I was counting on not getting that many in my whole life." Ashton Kutcher and other celebrities tweeted about the video, and it was mentioned on TV shows like *The Soup* and *The Daily Show with Jon Stewart*. "It just started spiraling. It was in the millions. And I didn't know my butt from third base about how to negotiate any of this stuff or protect myself legally. [My video] was on the internet, so everybody thought it was free," Schmidt said.

To help sort through the chaos, Schmidt reached out to his best friend's son, Ben Lashes. Lashes previously led a popular band called the Lashes, but had recently given up performing to work deeper in the entertainment industry. Schmidt sought him out for advice on his feline ingénue.

When the pair connected on the phone, Lashes congratulated him. "I was like, Keyboard Cat is all over the place," said Lashes. "You must be rolling in dough at this point!" Schmidt replied that in fact he wasn't. YouTube hadn't paid him a dime because he wasn't yet enrolled in the Partner Program. Later, O'Farrell said he only made $500 in ad revenue off of the clip. As Schmidt, Bahner, and others discovered, the advertising opportunities at the time were paltry.

Keyboard Cat was everywhere. Shouldn't that amount to something? Lashes imagined what would have happened if Walt Disney put Mickey Mouse on YouTube. How would Disney have tried to grow and protect the brand? Lashes called up Hollywood agencies. "I was like, alright I've got this cat and it plays a keyboard and it's got millions of fans. There's a lot of opportunity here," Lashes said. "People thought I was pranking them. They'd hang up on me."

No one really understood how to value online attention and influence. Lashes wasn't dissuaded, however. He encouraged Schmidt to create more Keyboard Cat videos (with a new cat he adopted named Bento since Fatso had passed away). Schmidt followed his advice. His new videos got more views and his YouTube channel, more subscribers.

Microsoft reached out, looking to use Keyboard Cat in a marketing campaign. Still working his day job, Lashes stepped outside and negotiated the deal by phone. It was tens of thousands of dollars, nearly the amount of his annual salary.

"There was no stopping us then," Lashes said. "We kept doing more content and social media and really treating it as a character with a brand. My rule with brands or anyone who came to us was: Respect the cat, and as long as you respect the cat we're good."

Schmidt and Lashes signed more deals. They launched merchandise and monetized in nearly every way they could. Shortly after the Microsoft partnership, they were contacted by Wonderful Pistachios. Schmidt initially told Lashes that the company offered him $800 to feature Keyboard Cat in a commercial. "I sent Ben [the offer] and said, 'Look, they said that I could be in their commercial for free, and then I'll be famous!'" Schmidt remembered.

Lashes, however, had other ideas. He rejected the idea that talent from YouTube was worth less than traditional talent. He went back to the negotiating table. By the time they hammered things out with Wonderful Pistachios, Schmidt and his cat received an all-expenses-paid trip to Hollywood to shoot "Keyboard Cat's Wonderful Pistachios." The commercial aired for five years, and Schmidt ended up making over $120,000 for the campaign.

The Wonderful Pistachios experience was further proof that online creators could earn big money, provided they were well-managed. Lashes had brought on Kia Kamran, a Hollywood lawyer who was among the first in the industry to take the intellectual property of viral stars seriously. Kamran chased down unauthorized Keyboard Cat merchandise and started vetting all contracts.

Lashes's success with Keyboard Cat made him the go-to manager for early viral internet stars. Before long, he quit his day job to become a full-time meme manager. He proved a savant when it came to spotting viral internet figures early and growing them into multinational consumer brands—Grumpy Cat, Nyan Cat, Success Kid, Ridiculously Photogenic Guy, Scumbag Steve: think of any big meme in the 2010s and, chances are, Lashes was their manager. He not only signed them but also helped each creator make the most of their digital fame.

Opportunities for viral stars expanded exponentially in the early 2010s. "Things were already starting to grow faster and become more mainstream," Lashes noted. It was in no small part thanks to him and other smart, hands-on managers. And YouTube was paying attention.

※

A growing number of YouTubers received millions of views and amassed hundreds of thousands of subscribers. Viewers were hungry for more videos, but producing regular content could be a full-time job. How long could creators keep doing that for free? And why should they?

At the time, the concept of paying users was radical. Just one online video competitor, Revver, shared ad revenue with users, and it had a fraction of YouTube's market share. YouTube didn't have much revenue at all, and it was racking up expenses fending off lawsuits and hosting millions of large video files. To surrender a chunk of its own earnings to pay the very users who were driving up those costs seemed absurd to many in Silicon Valley.

YouTube's decision came down to its biggest challenge in its early years: advertising. The company's only realistic path to long-term profitability was through ad deals. The biggest advertisers, however, were hesitant. Brands were used to running ads on TV, radio, and in print, where they knew what programming their ads would appear alongside. On YouTube, brand executives had no idea where their ads were going. They shuddered as they pictured their logos alongside pirated movies or high school potty humor.

Eric Schmidt sent a clear message to the team: in order to make ads work, YouTube needed premium content. Most of Google thought that meant big-name licensing deals, focusing on how to get TV clips on YouTube. But as Strompolos himself looked across YouTube's network of "oddballs," he saw that young people were doing something novel. "They were using YouTube to tell their stories," he recalled. "They were building audiences, and they treated it like their own television channel. And more so than just TV, they were connecting with their audience and developing an actual friendship with their audience."

When Strompolos took the reins, he saw YouTube's new creator engagement program as an opportunity. "To me, the revolution was that all these new voices had the chance to emerge, but we had to make it sustainable. We had to make it possible for them to be rewarded for their creativity."

For Strompolos and Hurley, there was another path to success. User-generated content had driven the early growth of YouTube. Could it also be its long-term strategy? What if user-generated content could be their premium content? If users weren't up to creating that type of content, then maybe the company could help them get there.

YouTube leadership approved of this plan. Thus, the YouTube Partner Program was born. YouTube would share ad revenue with creators, which would give them an incentive to keep making videos and the resources to produce higher-quality content (at least by Madison Avenue's standards). As their videos improved and gained more views, they'd become more desirable and consistent for top advertisers. Helping creators monetize created a virtuous cycle, where everyone would win.

Many key questions remained unanswered, recalled Strompolos. "Who should be eligible to monetize? How would we serve ads from Google AdSense onto YouTube? What would we do if people uploaded things they didn't own? How big did we want this program to be?"

By 2007, they were ready to onboard the first, handpicked group of YouTubers to the program. "We'd say, hey, we're the people from YouTube and we're super impressed that you've been creating videos," recalled Strompolos, "we want you to keep doing that, so we've created this program where you can have a share of the ad revenue!" Strompolos spent hours walking the chosen creators through how the program worked. While the novelty of the program perplexed some users, the notion of a potential payout for their creative work was enticing. Many early YouTubers were college students or young people without financial stability. They were elated at the chance to earn even a few hundred dollars for something they'd previously been doing for free.

The YouTube Partner Program launched in beta in May 2007 with a handful of content creators to start. They included Lonelygirl15; a popular skit-comedy duo, Smosh; and Lisa Donovan, aka LisaNOVA, a creator who posted skits and parody clips.

Throughout the rest of 2007, Strompolos and his team onboarded about ten to fifteen new creators per week into the Partner Program beta, including Tay Zonday. In December 2007, the Partner Program launched publicly, and any creator could apply to join. Strompolos continued to work as its manager, effectively overseeing the emergence of the first major class of YouTubers.

"Today when you think of YouTube, you think of creators," he said, "but in the early days that wasn't the case. We had to really champion this idea that there was this new class of people. They found YouTube, and it was their home, and we were there to help them thrive."

The Partner Program ran on a 55–45 split, where creators received 55 percent of the ad revenue on their videos and YouTube kept the rest. In the program's early days, the income from the program wasn't life-changing for anyone. Early YouTube was still an offbeat subculture that mainstream culture didn't yet acknowledge. To most creators, the platform was still a fun hobby, not a vehicle for fame or riches. But it was now the only major platform on the internet offering to pay users, an idea anathema to every other social media company on the rise.

※

YouTube's creator landscape metamorphosed in the years following the Partner Program's rollout. The platform's growth continued exponentially through the late 2000s, and the novelty of online attention was increasingly driving real business. A nascent industry was forming around creators who went viral, to translate budding micro-fame into bona-fide fortune. But there was still much to be figured out.

When Adam Bahner moved to Los Angeles in 2008 to make good on his viral stardom, it was a gamble. As a member of YouTube's Partner Program, he was now earning an income off his work, but there wasn't a clear path for viral creators beyond that.

The center of gravity for YouTubers was on the west side of Los Angeles, in Venice Beach. Bahner settled into an apartment in L.A. and began creating more music videos. He wasn't going as viral with his follow-up videos, but he had a subscriber base in the tens of thousands and took various gigs to make ends meet.

Those gigs were not the kind of work he'd hoped for, to say the least. "None of the traditional institutions cared that I had views on the internet," he said. "I'd talk to traditional agents in 2008 and they'd pick me up like a wet piece of lettuce and be like, 'where are your TV credits, where are your movie credits?'"

The highlight of Bahner's year came toward the end of 2008, when he hosted YouTube Live in San Francisco, a big event that Google threw to highlight YouTube creators. Among the performers were Akon, Katy

Perry, and Will.i.am, along with musically inclined YouTube creators like Bo Burnham and the Gregory Brothers.

The event galvanized YouTubers in Los Angeles. It was the first time many had been able to hang out in real life and they bonded over the similar challenges they faced. The king and queen of the YouTuber social scene in those days were a couple named Lisa Donovan (LisaNOVA) and Danny Zappin. Bahner rode with them back to the airport after YouTube Live. The YouTube Partner Program hadn't meaningfully benefitted Donovan, and, like many YouTubers, she and Zappin were growing impatient. They wanted their big break in Hollywood.

As they drove, Zappin vented to Bahner, "I'm so sick of Hollywood screwing over YouTubers. I'm going to make a studio just for YouTube creators, so only YouTubers can decide the terms of their success." One early YouTuber, Kassem Gharaibeh, told *Business Insider*, "We needed more than what the YouTube Partner Program could provide."

Zappin and Donovan envisioned a United Artists for the digital era. A single organization that would give its members promotional, management, and ad-sales support, while helping YouTubers grow online and potentially generate mainstream Hollywood interest.

There was some precedent for such a venture. Back in 2006, Next New Networks was founded by two former MTV executives, Herb Scannell and Fred Seibert, along with Emil Rensing, Tim Shey, and Jed Simmons. Their idea—before the YouTube Partner Program—was to bundle online video into larger channels, like you'd find on TV, and sell ads against them. The company created a range of channels themed around topics like cars, fashion, and pop culture. For each channel, Next New helped produce original programming and filled out the rest of its programming slate with videos from contributors.

Next New Networks pitched itself to creators and advertisers as a reliable way to get noticed: "You contribute your own videos, your comments and ideas, and we pull it together with original content to deliver a regular and dependable experience for you and the people that share your interests," they declared on the company website.

Next New Networks believed that the future of advertising was in online video creators, and it thought it could solve some of YouTube's ad woes by luring advertisers with strategically packaged content. The company was "challenging the idea that the chaotic terrain of sites like YouTube and MySpace can't be a friendly place for advertisers," the *New York Times* reported. Next New Networks secured $8 million in seed capital in 2007 and another $15 million the following year.

The company signed creators who fit well with the channels it already had in mind, like automotive and technology. It also used its funding to court independent video creators and produce shows that rose above the average production value of online video at the time.

In 2009, before other companies began sinking money into web series, Next New helped develop the hit "Auto-Tune the News" series with the Gregory Brothers, a team of siblings who created music parody videos. The series became an online sensation. Next New also inked a deal with Julia Allison to launch a style network and host a YouTube show called *TMI Weekly*.

Other creators in the Next New universe included people like Michael Stevens (known online as Vsauce) and Ben Relles, creator of the Obama Girl videos. "Our homegrown group of creators were all on top of one another in our studio, making videos and appearing in each others' shows," said Tim Shey, co-founder of Next New.

Next New threw weekly parties with Tumblr in their shared office above Les Halles, the French restaurant where Anthony Bourdain cooked, often packed with early bloggers and web video creators from *Gawker*, the *Huffington Post*, BuzzFeed, and College Humor. "The kids in the Next New universe—like kids of hipster scenes immemorial—dress a certain way, talk a certain way, manifest a certain cosmology," the *New York Times* declared that year.

"I wanted Next New to feel like this little world of renegade TV networks made by creators, and we fully embraced a low-budget, punk rock aesthetic for all of our shows and videos," Shey said. "Rather than having all of sunny L.A. to shoot, we had a warren of cramped, tiny green-screen studios on the twenty-first floor of an office building in NYC."

Donovan and Zappin had a different strategy. They started by establishing the first major content creator "collab" channel on YouTube called the Station. At the time, the concept of a collab group—an online super-group that could boost the profiles of each of its members—was revolutionary.

The two recruited many of the major content creators on YouTube at the time. Aside from Zappin and Donovan, there were eight other members: Dave Days, Philip DeFranco, ShayCarl, Shane Dawson, TheBdonski, WhataDayDerek, Rawn Erickson, and Kassem Gharaibeh.

The Station collab channel was the first project that Zappin and Donovan created out of a new company they formed in early 2009, Maker Studios. Maker started with just a handful of employees, and Donovan went into $200,000 of personal debt to help finance the new company. It was a huge risk, especially because the 2009 economy was still stuck in a recession. A significant chunk of the initial money went into renting a home in Venice, allowing Shay Carl to move to L.A. with his kids. The house, at 419 Grand Boulevard, became known as the Station, the first ever social media creator content house.

The Station became the fastest-growing YouTube channel in the site's history, shooting to the top of the most-subscribed channel list even before the group posted a single video. It stayed there as the team got to work. Soon the members of the Station were posting one video a week to the collab channel, usually an original sketch comedy bit. As the group settled in, 419 Grand Blvd also became the epicenter of the creator ecosystem. The *Hollywood Reporter* dubbed it "a mecca for the YouTube creator-community, a digital media version of 1960s Haight-Ashbury or the Sunset Strip."

The home effectively had an open-door policy for early content creators, as long as you knew one member of the Station. (Among the YouTubers who regularly couch-surfed there were Dax Flame and Bo Burnham.) Many of the group's YouTube videos were filmed inside the house or in its backyard. "Every YouTube star you could think of came through that place and either partied or crashed," Zappin told the *Hollywood Reporter*. "Thousands of videos were made there. We eventually got rid of it because it was becoming a liability. Too many parties."

Everyone involved in the Station had equity in Maker Studios, but Zappin and Donovan were the ringleaders. Even though they were still in serious debt, they invested further in Maker Studios' production capabilities. They hired directors, editors, and producers to create more content and elevate production values.

Revenue started coming in. Not long after launching, Maker landed its first brand deal, a $60,000 contract with Sanyo to promote its new line of camcorders. This deal demonstrated that they were onto something. Within a few months, Maker was poised to prove just how powerful a medium YouTube had become.

※

George Strompolos followed Maker Studios' emergence closely over the course of 2009, working closely with creators at the Station. "By then, you had this idea of viral videos being synonymous with YouTube," he told me, "and this idea of online content creators was becoming more mainstream."

The turning point for advertising agencies came in mid-2009, when the fast-food chain Carl's Jr. purchased two large banner ads on the YouTube home page for several hundred thousand dollars a day. After buying the banner ads, Carl's Jr. still had $90,000 left in their budget, so they asked the Google ad sales team what else they could do.

Strompolos suggested giving nine YouTubers $10,000 each to produce a video about Carl's Jr. to post on the same day the banner ads went live. "By today's standards, that is really small, but in 2009 it was a huge deal," Strompolos explained. "They'd never gotten a big brand to sponsor them yet."

Strompolos reached out to several members of the Station, who gladly accepted the terms of the campaign. Each YouTuber created a video showing how they ate a Carl's Jr. burger, a twist on the brand's infamous commercial with Paris Hilton, where she showed how *she* ate a burger (impractically sexily, while wearing diamonds and washing a luxury car).

Ezra Cooperstein, VP and director of the ad agency running the Carl's Jr. YouTube campaign, told Adweek he was hoping the creators' videos would give the campaign a modest boost. As he saw it, the real campaign was their takeover of the YouTube home page.

Once the campaign launched, however, Cooperstein was stunned. The nine creator videos collectively amassed over 11 million organic views within forty-eight hours. Meanwhile, clicks on the home page banners generated barely 100,000 views of the brand's official commercial. What's more, the YouTubers' followers were far more engaged, staying with the content longer than those who clicked a home page banner ad, then quickly navigated away.

The campaign was a light-bulb moment for the online creator industry. Strompolos and Cooperstein both felt as if they had glimpsed the future. "It seemed like all the value was with these creators," said Strompolos, "and they were way more valuable than the world realized." Creators, too, were emboldened by the success, which was proof that the influence they'd cultivated online had real monetary value.

Over the course of 2009, the amount that advertisers spent on YouTube rose rapidly. To those who were watching closely—like Cooperstein, Strompolos, and the creators at Maker—the true potential still remained untapped. Many media observers, however, continued to see YouTube as Google's folly. *Business Insider* declared that "YouTube Is Doomed" at the end of 2009, noting that the year's $240 million of ad revenue fell massively short of its $711 million in operating costs. "Presumably, the videos YouTube is already monetizing represent the best content available," the article offered, "with diminishing returns as they reach deeper and deeper into a repository rife with copyright violation, the indecent, the uninteresting, and the unwatchable." Google had paid top dollar, in other words, for the internet's backwater.

Those outside media saw things differently. By the end of the year, Cooperstein decided to leave his marketing job to pursue a career in online content. "That [Carl's Jr.] campaign showed a lot of people who were paying attention that this online creator thing is more powerful than you realize," recalled Strompolos. "It was a pivotal moment in a lot of ways. That little ad campaign was a big deal for a lot of people." In December 2009, Ezra Cooperstein signed on with none other than Maker Studios, joining the company as its first CEO.

As Cooperstein settled in as CEO, Zappin and Donovan focused on building out Maker Studios' network of channels. Bundling YouTube creators together created a powerful network effect, allowing members to collectively bargain with brands, gain attention from YouTube, and source bigger deals. Maker made it its mission to scoop up all the top YouTube talent in L.A.

✳

Many others had their own version of Ezra Cooperstein's realization. Entertainment-savvy entrepreneurs recognized an opportunity to act as a liaison between the brands and creators. The first firms that did so on YouTube became known as "multi-channel networks," a term concocted by Next New.

Over the first few years of Next New's existence, its original concept—a network of theme-based channels devoted to subjects like fashion and cars—was successful enough. But in 2009, the company identified another advantage of its talent-bundling strategy. Subject-matter channels didn't grow nearly as fast as channels run by recognizable individuals or collectives. Video stars were developing cult fandoms. The kinds of collaborations that the Station had put together were dominating YouTube as a result.

Next New Networks changed its strategy to lean into this realization, launching a new program called Next New Creators in December 2009. The program was designed to discover and foster talent from across the web. Next New Networks would then partner with up-and-coming creators to distribute and monetize their shows. While similar in philosophy to YouTube's own Partner Program, it could offer more hands-on development and better resources than YouTube could at the time.

Those accepted into the Next New Creators Program received promotional support, editorial outreach, and a dedicated sales team. They also gained access to production facilities, staff, and training on best practices for growing their viewership.

The program solidified Next New Networks as a leader, along with Maker, among the new wave of companies that were emerging to service YouTubers. Just a few years earlier, Adam Bahner hadn't seen a path to follow toward online success. Now the options were multiplying, and online video makers were rapidly professionalizing. The Next New Creators Program signaled another important first: it pioneered the term "creator" to refer to the new class of power-users who were reshaping YouTube. No longer were they relegated to the realm of hobbyists and amateurs. They were creators, building something real and formidable.

// **CHAPTER 6** //

Creators Break Through

WHILE THE MYTHOLOGY AROUND SILICON VALLEY FEATURED YOUNG men who could see the future better than everyone else, what the rise of social media thus far had proven was that nearly all of those young men had been wrong. They each built a platform with the confidence that it would do one thing better than anyone else, only to be redirected and rescued by a community of creative users. As the number of users on social platforms grew, so did the influence of their largest creators. Creators began to exert an impact on mainstream culture and amassed fandoms on the internet that rivaled traditional fame. Big brands started paying attention. YouTube had established itself as an entertainment force to be reckoned with. Its first wave of stars were starting to see their efforts pay off. Hollywood agents were returning their calls. Deals started flowing.

This massive shift was palpable at the first ever VidCon in the summer of 2010. Two of the most popular YouTube creators at the time, the "Vlogbrothers" Hank and John Green, organized the three-day meet-up to celebrate online video creation.

Previous YouTube meet-ups like the 7/7/7 meet-up in 2007 were all about avid users connecting with one another. VidCon, on the other hand, was about online fandom. It had become apparent that online audiences dreamed of meeting creators. The Green brothers decided to create a space where fans of theirs and other early YouTubers could come together and celebrate the communities they'd built together.

"We wanted to make VidCon reflect the astonishing diversity of online video itself: While the prevailing opinion within old media is that online video is primarily about sneezing pandas, we know that the true story is much more complicated," the VidCon program read.

The proceedings were bumpy. Microphones malfunctioned, cameras were lost, and lanyards ran out. The event felt very DIY. Hank Green ran around with his laptop open, attempting to stream from it. But attendees didn't care about the production hiccups. It was just a rush to be there, seeing in person the people they so loved watching on screen.

Some YouTube stars were shocked that anyone showed up at all. They were also shocked by the age of their fans. YouTubers themselves were almost universally in their twenties or early thirties, but fans attending VidCon included plenty of high schoolers. "I think everyone was surprised that the demographic that showed up was so young," said Michael Gregory, one of the Gregory Brothers. "It was adjacent to teen stars."

When one fan handed Hank Green a hand-knit stuffed anglerfish, a reference to one of his videos, he became overwhelmed. "I don't know what it was about that experience," he later told the *Hollywood Reporter*, "but I had to go hide and cry. It really changed my perception of what I was doing."

The barrier between fans and stars was also porous in a way that it isn't anymore. The fame achieved by early content creators was a kind of fame that kept them down to earth. YouTubers at the first VidCon were friendly and accessible. "It was this very small thing," said Gregory. "One night everyone performed, then you'd hang out on the couch. There was a lot of ukulele music going on. Everyone hung out with each other."

For fans and creators alike, the energy of the event was magnetic. It was the first time many creators had met their audience in person. The YouTuber Charlie McDonnell was swarmed by dozens of fans. Groups of attendees would nervously huddle around other YouTubers and ask for photos with their digital cameras. Many fans also had their own

YouTube channels, and here was a chance to learn from the big shots directly. Tyler Oakley attended the first VidCon as a fan, only to become one of the biggest faces of YouTube just a couple years later.

At the height of the conference, members of the Station performed a parody rendition of Justin Bieber and Sean Kingston's hit song "Eenie Meenie" as the crowd cheered. John Green danced onstage in a turkey costume. Another friend was dressed as Batman. During a Q&A that followed, members of the group spoke about collaboration being the key to keeping YouTube alive. They said the whole event was a testament to the power of that collaboration.

It was a wholesome, unabashed moment, but already it had a bittersweet tinge. The Station was fracturing, as some of the group's breakout stars ventured out on their own. Shane Dawson and Phil DeFranco both left and found success as independent creators. "We were all becoming more and more famous every day and making a lot more money every day," said Kassem Gharaibeh. "So we were suddenly dealing with other people's egos, our own egos, money, and trying to grow the studio."

※

YouTube staffers at VidCon nearly matched the fans' enthusiasm. YouTube director of product management Hunter Walk said he was blown away by "the two-way interactions between YouTube staff and community, the chance to go beyond the username and talk with people about what they need from YouTube's product to help them accomplish their goals." Falon Fatemi, a business development strategist at YouTube, wrote, "Although it was a bit uncomfortable to hear a 13-year-old girl ask Shane Dawson to be the father of her children, it was truly inspiring to witness the enthusiasm, dedication, and excitement of fans and YouTube stars."

The company used the event to announce the expansion of its creator support with a new $5 million Partner Grant Program. George Strompolos was aware that the Partner Program had created a chicken-and-egg scenario, where creators needed money to make the videos that would then allow them to raise money for subsequent productions. As he explained, the funds would serve as an advance against the partner's future YouTube revenue share. The Partner Grant Program distributed grants to partners and ended the year by doling out $1,000 credits at video-equipment company B&H Photo Video to five hundred top partners.

By the end of 2010, YouTube had enrolled more than 15,000 creators in its Partner Program. Strompolos's division reported at the time that the number of partners making over $1,000 a month shot up by 300 percent. Hundreds of creators were earning six-figure incomes. In the summer of 2010, Next New Networks' Gregory Brothers released the "Bed Intruder Song," an auto-tuned remix of the viral Antoine Dodson "bed intruder" clip from a local news segment. Their single landed on *Billboard*'s "Hot 100" chart alongside singles by Katy Perry and Usher. The song sold nearly 100,000 copies on iTunes, and its video earned over 20 million views by the end of August.

In 2009, fifteen-year-old Nebraskan Lucas Cruikshank, who portrayed an alter ego named Fred Figglehorn on YouTube, became the first YouTube creator to hit 1 million subscribers. Cruikshank was raking in tens of thousands of dollars a month from the platform, and his YouTube fame earned him a guest spot on the Nickelodeon sitcom *iCarly*. In 2010, he went further, with *Fred: The Movie*, released on September 18. It was the first Hollywood production featuring a YouTube star, alongside actors Siobhan Fallon Hogan and John Cena as Fred's parents.

By 2010, more creators were crossing that 1 million subscriber threshold. Among them was Annoying Orange, a channel that featured a series of personified fruits created by filmmaker Dane Boedigheimer. Annoying Orange garnered close to half a billion views and over 1.6 million subscribers, making it the eighth most popular YouTube channel of all time. By the end of the year, Boedigheimer was shooting a pilot and in talks with Cartoon Network.

These encounters with Hollywood reinforced the value of YouTube's openness. When a creator made something, they could post it without hassle. Once they started dealing with a professional studio, even to make a modestly budgeted movie like *Fred*, it was impossible to retain creative control. Cruikshank didn't write, cast, or direct *Fred*. The resulting film, which was not released in theaters and instead went straight to TV, was a disaster, earning the lowest rating possible from Rotten Tomatoes, a 0.0. The reason YouTube creators flourished on YouTube was that they could avoid the suffocating constraints imposed by gatekeepers. Even with millions of fans online, it wasn't clear if they could flourish offline.

※

The early 2010s were a boom time for companies like Maker Studios and Next New Networks. Known as "multichannel networks," or MCNs, they would bundle large YouTube creators together to collectively negotiate brand deals and business opportunities. MCNs offered production and editing services so that video-makers could focus on creating. At the same time, MCNs offered creators 24/7 technical support and liaised directly with YouTube on their behalf. In return for those services, creators would share a percentage of their revenue, often around 20 percent.

As the business model for MCNs solidified, they invited a new wave of investment. "There's a turn in the tide happening in the world of web video," Tubefilter reported at the time. "The second half of 2010 signaled a resurgence of investor interest in the business of making original programming for the internet. And those investors are once again bullish on original content."

Maker Studios raised $1.5 million in funding in December 2010 and another $1.5 million in early 2011, led by Greycroft and GRP Partners. Comedy network My Damn Channel secured a $4.4 million investment. Gaming network Machinima raised $9 million in the summer of 2010. Serious money began to appear in the YouTube ecosystem.

For years, Strompolos had watched mainstream brands dismiss YouTubers. "I would go to meetings with our sales team," he said. "I remember being in a meeting with Revlon, and they were telling me the economy was down, it was 2009, and they said, 'Hey, when the economy is down, we actually do really well selling our low-cost products like eye shadow and lipstick, because it's a cheap pick-me-up for people.'"

Strompolos got an idea: At that time, the number one female YouTuber was a woman named Michelle Phan, and videos about eye shadow and eye makeup were a key part of her content. But when Strompolos told this to the Revlon executives, it didn't go over well. "So imagine me, the YouTube guy, sitting next to a Revlon executive, hearing her tell me this. I blurted out all this cool stuff and, to my surprise, she laughed. She said, 'Oh, these kids. Where do they find the time? Now, that's not for us.'"

Nothing ended up coming out of the meeting. Michele Phan went on to build a $500 million company and start her own beauty line without Revlon.

Strompolos kept having moments like this. "I would then go to Hollywood and I would talk to the agents and the studios and the networks and I would say, 'Hey, there's a new generation of creators. They've captured the hearts and minds of young people. And this is the future of entertainment. These are the new voices.' My whole speech. And they would say, 'That's low-quality user-generated content. That's not what we do. We do premium shows and movies.'"

Eventually, Strompolos became tired of the disconnect. "I knew that this revolution was happening. I was fortunate to be at the epicenter of it all. And I said, 'You know what? I think this change in what's possible now is so profound that new media companies will be born.' And I remember an aha moment. I was flying between SF and NY, just thinking and I had that thought, 'New companies are going to be born. Which one of my friends are going to do that?' And I said, 'Wait a second, I should do that.' And I had one of those moments where I'm like, 'Get me off this plane!'"

By the end of 2010, Strompolos had quit YouTube. He was friendly with Danny Zappin and the Maker Studios people by this point, so he moved into Zappin and Donovan's guest house in Venice Beach to focus on his next project.

Strompolos knew that Maker Studios had already signed nearly all the top L.A.-based creators. The largest MCN at the time, Machinima, had cornered the market on gaming. The Collective, a management company founded by Michael Green, had begun working with YouTube personalities like Fred and Annoying Orange under its Collective Digital Studio venture. Collective Digital Studio eventually became a leading MCN under the name Studio71. Other MCNs were cropping up focused on different niches like comedy and prank creators. But nearly all were focused on recruiting the biggest talent.

Strompolos was always most interested in the talent emerging from the ground up. He saw his big opportunity to be working with the next generation of creators, the mid-level and rising YouTubers. He took a more technical-forward approach, building tools for sponsorship integrations, managing collaborations, and streamlining copyrights and royalties.

In 2011, he launched Fullscreen, his own MCN. He brought onboard his previous collaborator, Ezra Cooperstein, who left Maker to join on as Fullscreen's COO. With Strompolos's experience, Fullscreen became a major player in the space overnight, becoming the top MCN alongside Maker Studios.

Creators who joined the Fullscreen network could log into the company's system and access a variety of tools to grow their audience, engage with fans, and track how their channels were performing. Fullscreen's dashboard also showed them how much money they were generating and offered recommendations for what they should do to optimize their content.

Like other MCNs at the time, Fullscreen quickly took on venture capital and joined the race for additional funding and talent. By 2011 Maker represented over 150 channels, generating 325 million views per month. Malik Ducard, former director of content partnerships at Google who ran a team that managed YouTube's ties to the MCNs, said he saw the companies as an essential element of the platform's ecosystem. "So much innovation comes from this group, and this group often pushes the envelope for YouTube in a lot of ways," Ducard said at the time.

YouTube realized that it couldn't fall too far behind MCNs in the services they offered creators and advertisers. If they did, they could be cut out of their only source of revenue, advertising. So in March 2011, YouTube announced that it was acquiring the original MCN, Next New Networks.

Next New Networks had already been acting as a go-between for creators and YouTube for years, and now, as part of YouTube, it brought that team in-house. As Next New Networks was integrated, it merged with other areas of YouTube's business to form a new team called YouTube Next, which would eventually become the YouTube Creators Program. Tim Shey, Jed Simmons, and Lance Podell from the Next New management team launched the YouTube Next Lab. Next New co-founder and CEO Fred Seibert resumed working as an independent producer at his production house, Frederator Studios. Seibert was named as a new YouTube channel partner in November 2011.

V Pappas, who also moved over from Next New Networks, became YouTube's global head of audience development. "V led the creation of the YouTube Creator Playbook that pretty much revolutionized the industry," Shey said. "It was the first time anyone had written down all these principles, like publishing regularly, optimizing your videos for search. . . . It was all the tips and tricks to being a successful creator." Pappas went on to become chief operating officer at TikTok.

In the coming years, the YouTube Next team spearheaded a series of initiatives similar to the Partner Grants Program, which had proven a massive success in 2010. The YouTube Next team expanded YouTube's presence at offline events like VidCon and Playlist Live. They were tasked with the now-familiar goal of making YouTube a place for high-quality content, which they hoped premium advertisers would pay serious money for.

The team also finally did away with the "partners" label. "Partner" was misleading now that there were so many in play. At the same time, no one wanted to be pigeonholed as just a "YouTuber" or "YouTube star." So they adopted the term that Shey and his colleagues coined for Next New's program: creator.

"These people were more than on-screen talent," Shey told me. "We thought they were amazing. They all had the full-stack skill set. They could write, edit, produce, do community management, and were entrepreneurs. The phrase YouTube star didn't capture everything they did and was used in a derogatory way. It wasn't considered legitimate back then."

So, YouTube "partners" became "creators." YouTube was so successful at pushing the term that the whole tech industry adopted it. In the meantime, the company proceeded to build an entire infrastructure around them.

※

One of the earliest attempts to validate and quantify online influence was a platform called Klout, which launched in 2008. The app used social media analytics to assign users a rating called a "Klout Score," a number between 1 and 100, 100 being the most influential. It took into account signals such as a user's follower count, post frequency, and topics of interest. Brands could reward users who were highly influential on certain topics with freebies and gift cards.

The tech press balked at the idea and the platform was roundly mocked. Seeking to leverage influence on the internet, or even acknowledging its existence through a score, was considered crass. But while older people laughed, younger users embraced Klout scores, albeit somewhat ironically. They loved the idea that they could get free stuff just for posting a lot on the internet.

In 2011, a group of eight college students in their early twenties formed the "Yung Klout Gang." They posted relentlessly, primarily on Twitter, and used the internet to build up their social status and land jobs in the music industry. Fans even set up a Tumblr page to catalog the group's escapades. But despite their growing influence, there was still not much Yung Klout Gang could do besides collect gift cards. In 2013, MTV approached the group about a reality show, but they declined. "At that point, there was no influencer life. What would we have done?" said Lina Abascal, one of the group's members told me. "Be embarrassing on MTV? You wouldn't even get 1 million followers from that back then."

But the birth of Klout signified a growing awareness of metrics, and average users began to recognize online influence as valuable, even if it wasn't totally clear how it should be quantified. People were beginning to become aware of their own social rankings, and they watched big creators' numbers climb higher and higher.

※

In September 2012, photos of Tardar Sauce, a female blue-eyed mixed-breed cat with a distinct frowning expression, were shared to the Reddit board r/pics. The photos of her immediately gained traction, racking up over 25,300 upvotes in less than twenty-four hours. The same photos of Tardar Sauce posted on the photo sharing site Imgur garnered over 1 million views in under two days.

Tardar Sauce's unique "grumpy" expression, caused by underbite and feline dwarfism, quickly became a meme. She became known online as "Grumpy Cat." Overnight, she was everywhere.

Ben Lashes, watching Grumpy Cat's rise in amazement, got in touch with Tardar Sauce's owner, Tabatha Bundesen, and her brother, Bryan Bundesen, who had posted the photos on Reddit. Lashes and the Bundesens began working together and one of the first things Lashes did was get Grumpy Cat featured on the *Today* show. The group met up the night before Grumpy Cat's TV debut, a few blocks from Times Square in the fall of 2012.

People began stopping the Bundesens and Grumpy Cat on the street to take photos. Even those who didn't know about the animal from the meme were smitten.

"When [Tabatha] pulled out Grumpy Cat it was like angels from heaven were singing," Lashes told me. "She looked better than the photos and videos. She was the coolest, chillest cat everyone falls in love with immediately. It was just this magnetic force. She had a power on people even if you didn't know her from the internet." Lashes was, he realized, in the presence of a star.

"I was like, 'Man, I've been saying for years that there's going to be something from the internet that is the viral entity that can come and compete with the mainstream entertainment industry,'" recalled Lashes. "There were bits of that with my early [meme] clients, but none of them had been able to compete all the way. With Grumpy Cat, I was like, 'This is the chosen one. It's going to change things going forward for the internet and pop culture.'"

Shortly after signing Grumpy Cat, Lashes landed the feline's first big deal, a partnership with Purina cat food. Lashes flew with the Bundesens to Austin, Texas, in March 2013 to shoot the campaign. Later that month, Grumpy Cat was an official guest at the media company Mashable's South by Southwest (SXSW) house. Mashable sponsored a Grumpy Cat photo area where people could snap selfies with the viral star.

Though the photo booth didn't open until one p.m., the line to meet Grumpy Cat began forming at six in the morning. Hundreds of people waited in the hot Texas sun for hours. Some brought snacks and folding chairs, all for the chance of taking a photo with Grumpy Cat. The line was over six blocks long.

"It's hard to overstate how big the hype was around [Grumpy Cat] that year," said Lashes. "It was like Beatlemania." National media lauded Grumpy Cat as "The Biggest Celebrity at SXSW." "Forget Elon Musk or Al Gore," CNN wrote, "the biggest star of the South by Southwest Interactive festival is less than a year old, sleeps all day and looks like she just swallowed a hairball. Meet Grumpy Cat." The same article noted, "Internet fame is becoming as potent a force as other types of celebrity."

Lashes helped transform what had started as a meme into a full-fledged brand, pioneering a revenue model for viral animals that many pet owners still follow to this day. Beside the YouTube channel, Lashes helped grow the Grumpy Cat website (which in 2013 attracted over 1.5 million unique visitors per month), a Facebook page, and later, Twitter and Instagram accounts. Grumpy Cat's face was plastered on millions of bags of cat food in pet stores across the country.

In 2014, the cat starred in a Grumpy Cat Lifetime Christmas movie, released worldwide with Aubrey Plaza as the voice of Grumpy Cat. Grumpy Cat appeared on *Anderson Cooper 360*, *Good Morning America*, *Fox and Friends*, and *American Idol*. She attended the MTV Movie Awards and visited the offices of *Vogue* and SiriusXM. In 2016, she joined the cast of the Broadway musical *Cats*. Lashes signed deal after deal.

"Grumpy Cat brought on that era of where anyone, or anything, can be an influencer," he noted. "Anyone with the creativity and drive could potentially make something go viral."

"Grumpy really broke a barrier of internet culture becoming pop culture," he concluded. "The cat helped a mainstream audience grasp the concept of memes and recognize that things created online could grow into successful mini empires."

Grumpy Cat passed away in 2019 at the age of seven following a urinary tract infection. In true celebrity fashion, her brand continues to live on. Grumpy Cat's face still appears in ads around the world. Her Instagram account is still dutifully posting to more than 2.6 million followers, advertising things like Grumpy Cat–themed Halloween decor, and hundreds of products are still for sale on the official Grumpy Cat website. Grumpy Cat slot machines recently started appearing at casinos in Las Vegas, and a Grumpy Cat cartoon series is in development.

A new era had arrived. When George Strompolos originally conceived the YouTube Partner Program, it was absurd to imagine that one day a whole industry would be built around such creators. In fact, he had posited something even more outlandish: creators would be more valuable to YouTube than TV and music licensing.

Even by 2011, when Strompolos left YouTube, that future was already in sight. He later said that at that time about 40 percent of YouTube revenue was generated through the Partner Program, which amounted to roughly $400 million. The numbers only rose. By 2021, YouTube reported $28.8 billion in ad revenue, and it passed along roughly $15 billion to creators. It was a staggering number out of a program that began with a five-person team trying to convince a handful of vloggers to take a leap.

PART III

New Dynamics

// **CHAPTER 7** //

Twitter Follows Back

The rise of digital influencers was not just a story of software but of hardware. Through the mid-2000s, blogs, news sites, online shopping, and social networks were the province of desktop and laptop computers. That would change in June 2007.

The iPhone untethered the internet from the desktop. The energy of the blogosphere was redirected toward faster, mobile mediums. Twitter, Tumblr, and Instagram facilitated the transition to mobile. These sites had learned from Facebook's streamlined profiles and easy feeds. But they diverged from Facebook's friend-focused model, favoring an open network that allowed users to "follow" anyone they found interesting. This mirrored the subscriber-based model of YouTube as well as the open architecture of the blogosphere. Instead of trying to re-create real-world friend networks online, this new generation of social apps sought to build an audience of friends and strangers alike. At the same time, they took the lessons of the blog era—chiefly, that anyone could build a following online—and expanded upon them.

As companies like Twitter and Instagram created intriguing new spaces online, they scrambled to support the flood of new users. It was the same problem that had downed some of their predecessors. (Remember those Friendster load times?) These start-ups raced to handle rapid growth and the technological demands that came with it. They had little bandwidth to appreciate and address the multifaceted ways in which users were making the most of their new digital tools.

※

Twitter was launched in 2006 by Jack Dorsey, Evan "Ev" Williams, Biz Stone, and Noah Glass. The product was in some ways a throwback to a previous internet era. Dorsey had adored AOL Instant Messenger (AIM) in the late 1990s, as well as an early blogging service called LiveJournal. On AIM, users could customize their "Away" messages, explaining why they had stepped away from the keyboard. On LiveJournal, bloggers could display a status message that told readers what they were up to, whether it was reading, cooking, coding, or walking the dog. While AIM and LiveJournal had fallen out of widespread use by 2005, Dorsey was still taken with the "status" feature.

At the time, he and his co-founders were struggling to find traction for their podcast product, Odeo. When Apple released its iTunes podcast app, the company found itself in a scramble for new ideas—*any* idea. During a last-ditch hack-a-thon, Dorsey and a few other developers threw together the code for Dorsey's "status concept."

Twitter's four co-founders each added their own ideas to the product as it came together. Stone wanted the text-based product to work on mobile phones, even before smartphones had rolled out. Glass drove much of the enthusiasm for the new idea within Odeo and ultimately came up with the name, Twitter (or "Twttr" as it was styled in the early days). Williams, previously the founder of Blogger, lobbied for Twitter to allow public posts as an option, so that it could serve as a form of microblogging.

In March 2006, the prototype of Twitter was ready to go live. It was "crude and simple," as journalist Nick Bilton described in his book *Hatching Twitter*. "'What's your status?' sat at the top with a rectangular box below, then an Update button that allowed people to share their status. Like a blog, a stream of updates flowed below."

On March 21, 2006, Dorsey posted the first tweet: "just setting up my twttr."

Early Twitter operated chiefly over the SMS text messaging protocol, which was originally why Twitter had a 160-character limit (it was later lowered to 140 characters to accommodate the twenty characters allotted to usernames). In its original pitch, Twitter described its purpose as a better way to keep up with friends. Twitter is "for staying in touch and keeping up with friends no matter where you are or what you're doing," its original website touted. Users could create a bare-bones profile and connect to friends. Then, they'd receive updates whenever those friends posted their latest statuses, or "tweeted."

In hindsight, the initial concept looks a lot like group texting—which didn't exist at the time—and that's what most people used Twitter for in its first two years. Most early tweets matched Dorsey's vision of simple status updates for friends. "Taking a walk to get a bus pass. Warm outside," Dorsey posted on January 1, 2007. "Randi and I just lip dubbed Like a Virgin in matching white suits. Hysterical :)" tweeted Julia Allison on April 26, 2008.

In their early years, Twitter and Facebook were often discussed in similar ways: both were technologies that opened up new forms of connecting with your friends. (Facebook's "status update" feature was rolled out just a few months after Twitter launched.)

But in a world that was already more connected than ever before, this created plenty of critics. "Twitter is a bad, bad thing," wrote *USA Today*'s tech columnist Andrew Kantor. "According to Twitter, you see, we should be in touch every second—every *moment*. This is madness." "Who really cares what I am doing, every hour of the day?" wrote *Boston Globe* columnist Alex Beam. "Even I don't care."

For many users, though, a switch flipped once they tried the service. Clive Thompson, in the *New York Times Magazine*, captured why Twitter felt so revelatory: "Each little update—each individual bit of social information—is insignificant on its own, even supremely mundane. But taken together, over time, the little snippets coalesce into a surprisingly sophisticated portrait of your friends' and family members' lives, like thousands of dots making a pointillist painting. This was never before possible, because in the real world, no friend would *bother* to call you up and detail the sandwiches she was eating." One Twitter user described the experience as "a type of E.S.P."—something that he wouldn't have known was possible, let alone desirable, before the dawn of Twitter. "It's like I can distantly read everyone's mind," the user told Thompson. "I love that. I feel like I'm getting to something raw about my friends. It's like I've got this heads-up display for them."

※

Twitter users soon realized that the company's service could be used for much more than the founders imagined. One of the most eye-opening moments came just a month after launch, during a mini earthquake in August 2006. Only a few hundred people were using Twitter at the time, mostly the company's employees and their friends. Dorsey was sitting at his desk one evening and felt the room rattle. The first thing he did was pick up his phone to send a tweet.

Before he could finish typing, though, he saw a tweet from Ev Williams: "Did anyone just feel that earthquake?" As Nick Bilton tells it, Dorsey dashed off his own status in response: "Just felt that earthquake. No one else here did." Moments later, "a string of other messages started streaming into his phone like letters falling through a mail slot to the floor. 'Agh earthquake,' wrote one friend. 'Yep. Felt the quake,' wrote another. Then a handful of other earthquake tweets appeared. 'I felt the earthquake but Livy didn't believe me until the twitters started rolling in,' wrote Biz."

It was a minor earthquake, but a major one too. Twitter had turned fleeting personal experiences into moments of connection. In this moment, Williams saw a vision of what Twitter could become: not just a way to share one's individual status but a way to share news and catch "a view into what was happening in the world."

Over the next year, as more and more users experienced their own earthquake moments, Twitter grew beyond the circle of the co-founders and their Silicon Valley niche. Its biggest spike came at South by Southwest 2007, when attendees signed up en masse, using Twitter to share the latest event buzz and highlights, as well as real-time intel on which party had the best free booze.

Over the course of 2007 and 2008, Twitter expanded to feature not just individuals and friend groups but accounts run by bloggers, companies, and organizations. The *New York Times* began using Twitter to relay headlines from its main site. Police and fire departments realized that they could use tweets to broadcast alerts in real time. Tech bloggers discovered that Twitter could be another forum for conversation about their posts.

At the same time, people set up parody Twitter accounts as fictional characters or celebrity figures. Darth Vader was a top Twitter account for a period, as was a (fake) Shaquille O'Neal. But then fiction became reality, as actual celebrities began to join, signaling an entirely new phase of the internet. Minor actors joined first, followed by ambitious politicians (among them a rising star named Barack Obama). In 2008, the real Shaq joined.

Still, no one at Twitter knew how to define what it *was*. "Everyone had a different answer," wrote Bilton. "'It's a social network.' 'It replaces text messages.' 'It's the new e-mail.' 'It's microblogging.' 'It's to update your status.'" As of 2007, no one even knew whether to call it a social network. Everyone had different ideas about what it should be. Dorsey saw Twitter as a service that allowed users to post updates about themselves and to follow their friends. Williams thought Twitter had proven that it was less about sharing personal updates and more about public conversation.

This wasn't an intellectual exercise: convincing companies to advertise was a lot harder when you weren't sure what you were doing.

The dispute eventually came to a head over a tiny but telling feature. Dorsey had set the prompt that users saw when they encountered the blank "tweet" box on Twitter's website as a personal, "What are you doing?" Williams thought that was too narrow. When he became CEO, he pulled rank and changed the prompt to "What's happening?"

The spat was representative of the ongoing chaos. Twitter rotated through CEOs, each with a different focus, none with a plausible business model. They were still figuring out how to define and pitch their service at a fundamental level. Meanwhile, the site kept absorbing new users and crashing often with spikes in traffic.

The internal conflict led to stasis when it came to Twitter's product itself. This may have been good in the long run: keeping the product experience consistent gave users time to discover how they wanted to use it. This allowed users, rather than staff, to determine what Twitter would become. The payoff, ultimately, was a new type of social network built on followers, not friends.

※

This iconic feature of Twitter—"followers"—now looks like a visionary choice, a phase change in the history of the internet. But it was far more pedestrian in practice. When Twitter initially launched, there were two forms of connection: "follow" and "friend." Users could optionally "follow" any of their friends, which meant they'd receive a text message whenever that friend posted. If you unfriended a friend, you would still see their tweets on Twitter.com, but you wouldn't get the texts.

Twitter worked this way for the first year. Each profile had a list of friends and a list of followers. Users, however, found the distinction between the two confusing. After a year of feedback, Twitter agreed to simplify things. By then it was clear that "following" seemed more appropriate, especially since plenty of users were "friending" blogs and news feeds and Darth Vader. So, in July 2007, Twitter nixed the friends list.

Intended to make the site easier to use, this change would set Twitter on a trajectory to become radically different from previous social networks, eventually more like YouTube than the rivals its founders were immediately focused on.

In that same formative period, some of Twitter's other iconic features also emerged. Twitter's founders had left no mechanism to sift through the growing stream of updates. As a result, users created their own conventions to help one another make sense of that flow: the hashtag, the @ symbol, and the retweet. Only later, after months or years, were official versions of the features incorporated into the Twitter product itself.

The hashtag, for instance, first appeared on Twitter in 2007, after it had been used on the photo-sharing site Flickr to tag groups of similar images. Twitter users started to use hashtags to keep track of related tweets, since they allowed for easily searchable strings of text. Open-source developer Chris Messina proposed in a blog post that Twitter could use hashtags to sort conversations into different "channels." In autumn 2007, when a wildfire sparked near San Diego, the hashtag #sandiegofire became one of the first instances of a widely used hashtag.

But when Messina pitched the Twitter team on an official hashtag feature, the co-founders pooh-poohed the idea. "Hashtags are for nerds," said Stone. They were "too harsh and no one is ever going to understand them," said Ev Williams. The co-founders suggested they'd "come up with something better later, something friendlier." But before they could get around to it, users embraced the hashtag.

Retweets followed a similar trajectory. They were first created manually by users, who would type "RT" and then copy-paste the text of a tweet to amplify it. Users were manually retweeting as early as 2007—bootstrapping a tool for spreading news and popular posts. Twitter didn't integrate retweeting into its product until late 2009.

Twitter's product development was slow, but its cultural impact was expanding faster than anyone had imagined.

※

In 2009, a passenger plane out of LaGuardia lost power and navigated a daring emergency landing in the Hudson River. In print journalism, the near-disaster was a front-page story the next day. Radio and TV stations reported on the landing as soon as they could. But it was a viral tweet posted mid-landing—not even afterward—by one of the rescuers that broke the news. The "Miracle on the Hudson," as the episode became known, marked Twitter's arrival as a cultural force.

A handful of stars had dipped their toes into the online world before, starting blogs or MySpace pages, but Twitter was the first social media site they widely embraced. The character limit made it easier for celebrities to write their own posts, and the directness of their posts was catnip for fans. Many celebrities had felt beholden to studios, record labels, or the paparazzi throughout the 1990s and 2000s. Suddenly they had a direct line to their audience. After actor Ashton Kutcher challenged and then defeated CNN in a race to 1 million followers, Ev Williams appeared on Oprah to celebrate—and help her post her first tweet. The rush of media coverage within just a few months had transformed Twitter from online novelty to hot commodity. Online observers and journalists declared it a watershed moment that marked the "changing of the guard: old media to new media," according to *CNET*. And Twitter soon proved that it had the power to elevate a new form of celebrity, something between the average user and traditional stars.

*

Aliza Licht, senior vice president of global communications for the high-end fashion brand Donna Karan, followed the shift to digital and personality-driven media closely. She had worked her way up the ranks of Donna Karan after joining DKNY, a sister brand, in 1998. By 2009, she was living the glamorous life of a fashion executive in New York City—though an exhausting one, too, thanks to a four-year-old and a one-year-old at home. Licht began hearing more and more about Twitter. "I was so obsessed with it," she later told me. She loved the idea of being able to speak to anyone around the world.

She had an idea. The DKNY brand had been working directly with fashion bloggers. As Licht watched the CNN versus Kutcher race to 1 million followers, she recognized that Twitter was quickly becoming a public forum where anyone could have a voice. She and her team wanted Donna Karan to be part of the online conversation and decided to create an account for the brand.

Fashion brands weren't yet on Twitter, and Licht worried that if she registered the handle @DonnaKaran, people would assume it was the founder tweeting. "I wanted to make sure Donna's voice was protected," Licht explained. "If people assumed it was her tweeting, it would be tough from a PR perspective."

In conversation with Donna Karan's marketing and public relations team, Licht suggested taking a cue from the hit show *Gossip Girl*. The show revolved around an anonymous commentator who documented the ins and outs of the lives of private school kids on New York's Upper East Side. What if, Licht wondered, the brand Donna Karan had its own in-house Gossip Girl?

"We thought about it," Licht said, "Everyone was like, 'What would this personality speak about?' I was like well we do a lot of fun things in public relations. We have lunches with editors, we go to fashion shows, we do celebrity dressing. We can do it all through the lens of PR, and name the account DKNY PR GIRL."

The team agreed and Licht got approval. Her legal team had one stipulation: that only one person have access to the account. Apparently they told Licht, "Since you're an executive at the company, you can be the only one who can tweet," Licht said. This stipulation would change Licht's career, opening the door for her to stumble into one of the first waves of social-media celebrity.

Licht logged into the DKNY PR GIRL account from her phone and began posting. Most brands had seen social media as free advertising, touting discounts and deals to redeem at physical stores. Licht, however, saw that when she tweeted as DKNY PR GIRL about a sale on handbags, no one liked or retweeted it, no matter how good the deal was. Instead, she learned that the key was to act human on social media. When she tweeted about celebrity dressing on the red carpet, or her daily routine, people engaged. Licht found herself creating an alter ego for the account based on her own personality. "Maybe she's getting a manicure, or she's getting hair done, or having lunch," Licht explained. "Whatever I was doing, the persona was doing."

Followers flooded in, first by the hundreds, and then by the thousands. DKNY PR GIRL became fashion's most prominent Twitter account.

The old guard, however, was not impressed. In fact, many were horrified. They saw the tweets about "Celeb X" (as Licht never named names) and DKNY PR GIRL's lunch dates as self-indulgent and cheap. It had taken years of work to build an elegant, luxury brand that consumers could aspire to own. To put a more human personality onto that brand (that wasn't the designer), they thought, was to devalue it.

The bigwigs weren't alone in panning DKNY PR GIRL. Someone at *DNR*, a fashion news publication, sent a scathing email to Licht's boss. Licht recalled, "They said they've been a longtime fan of Donna Karan, but how could the company let this 'PR Girl' talk about whatever she wanted on social media?" Luckily, she had enough clout at the company to stay the course.

Women's Wear Daily, on the other hand, praised the account. The fashion world took notice of the engagement metrics, and similar accounts popped up. Oscar de la Renta's communications head, Erika Bearman, developed a massive following under the handle @Oscar PRGirl. Throughout 2010, more brands developed personalities on Twitter. @Bergdorfs became a staple in online fashion circles thanks to its snappy commentary. Media companies hired their first crop of full-time social-media editors, primarily young Millennials whose entire job was to run brands' social media accounts.

On Twitter's end, the company was thrilled to see new accounts that boosted engagement, but it was also somewhat perplexed by the new turn in its user base. These accounts weren't individual users, media outlets, public figures, or even spokespeople. They were a strange mix of public and personal. The company's primary interaction with these early digital Twitter creators was to sell them on advertising on Twitter. While it had a celebrity partnerships team, the emerging creator class on Twitter was largely left alone.

Licht continued to run the account anonymously for a year and a half, until, in 2011, it started to become untenable. By then, DKNY PR GIRL was interacting with some of the famous celebrities and media figures who were flocking to the platform. Licht would retweet and reply to people like Chrissy Teigen and Shonda Rhimes. She live-tweeted commentary on shows like *Gossip Girl* and *Scandal*. Licht and the celebrities and media figures she befriended on Twitter began hosting "Twitter dinners," where they'd meet up IRL to dish about online life. More and more people were beginning to dig into who was actually behind the account. Licht and her colleagues at Donna Karan decided to come clean.

To announce the reveal, they shot a video over four days, documenting Licht at four Donna Karan fashion shows, revealing her as the real-life "DKNY PR girl." The post, which included the video, went viral on Twitter and spread far beyond the account's several hundred thousand followers. "The reveal got something like 230 million media impressions," said Licht. "There were headlines all over the world. I had no idea it was this big secret that so many cared about." It also immediately catapulted Licht into full influencer status, before that was a thing. Suddenly, she was everywhere. DKNY PR GIRL became a Halloween costume. She live tweeted the Met Gala, and *Vogue* streamed her tweets to its homepage. "DKNY PR GIRL became a whole course for college communications majors that people would study. There were no other examples at the time of someone who wasn't a designer or spokesperson creating a voice online that wasn't the brand," Licht said.

A 2012 *New York Times* article heralded Licht and the arrival of a new class of what the paper deemed "E-lebrities," which today we'd call content creators or influencers. Licht had forged for herself a distinct personal brand and had done so in a way that echoed Julia Allison's posting strategy. Since Licht's efforts were attached to a real, well-known brand, she largely received praise rather than disdain.

"Sure, it's a lot of fluff and a lot of noise," the *New York Times* declared, "but you could argue that Ms. Licht and Ms. Bearman have created platforms that influence the perception of their brands more broadly than the words of most critics. But their real impact, in 140 characters or less, is this: Other companies now realize that they, too, have to invest in social media, even if it's ultimately meaningless."

Of course, it wasn't meaningless. Twitter had created a new communications infrastructure that even its own founders had trouble defining. But its users had repeatedly shown the many ways in which social media could and would be applied.

What Licht and the first wave of Twitter stars discovered—and what even the Twitter co-founders didn't fully grasp—was that social media was practically a celebrity engine. It would change how fame and influence functioned.

The first round of top Twitter accounts started with some form of old-world clout: Ashton Kutcher was a TV star first; fashion companies had established brands. But what appealed about their particular online accounts was something that couldn't be gleaned from the old system—the dishy rapport, the direct access, the behind-the-scenes peeks free of intermediaries. Instead of paparazzi pics or an elegant fashion shoot, social media offered something more relatable—an immediate and intimate sense of personality, even if that personality was fictional.

The next step for social media was the influential personality unattached to an old-school brand or preexisting celebrity. Social media was heading toward that same idea that George Strompolos had grasped early at YouTube: soon you wouldn't need gatekeepers, old-school brands, or old-media celebrity to become an icon.

Aliza Licht realized this as well. She helmed the DKNY PR GIRL account for a few more years until stepping down in 2015. By that time, she'd become enough of a celebrity in her own right. She'd written a successful book and set out to build a business of her own, helping mentor younger women in the industry. That same year, Donna Karan herself stepped down from the label, and several other senior leaders left the company. To Licht, it felt like a good time to close that chapter of her life and offer the new company heads a fresh start.

Erika Bearman, @oscarprgirl, also stepped down from her position in 2015 to pursue consulting. Meanwhile, other fashion-industry figures who had amassed similar followings online began to venture out on their own. It was starting to become clear that influencing, if orchestrated well, could become a lucrative, full-time career. They didn't need a high position at a major brand anymore to succeed.

"Fashion was leading the way but at the end of the day everyone was recognizing this was the future," Licht noted.

✳

Without entirely knowing where it was headed, Twitter had succeeded in creating a single network that combined the online intimacy of Facebook with the creative energy of the blogosphere, all while adding an entirely new layer: the cultural sway of celebrities, blue-chip brands, and the political and media elite. Mark Zuckerberg had talked about connecting the world, but Twitter was the first to throw everyone—from DKNY PR GIRL to Barack Obama, from Shaq to fake Shaq, from your cousin to your coworker and future employer—into one roiling conversation.

As it did so, Twitter set a new standard for tech and media companies alike. Twitter's success would inspire every social media company that followed, and it would also force the reigning social network—Facebook—to change its model.

In fact, it sent Facebook scrambling. The company had practically invented social media, but Twitter revealed that some users wanted something more than friends. As Twitter became the darling of tech coverage throughout 2008 and 2009, Facebook knew it needed to respond. Its first strategy was to try to buy Twitter. When that failed, it settled for mimicry, even if that fundamentally changed Facebook's main offering.

Over the course of 2009, Facebook redesigned its site and rolled out new features in direct response to Twitter. Users could now become "fans" of pages created for celebrities, brands, or other entities, a mirror of Twitter's "follow" function. The company also reworked its News Feed to be more similar to Twitter's feed, intermixing updates from friends with updates from fan pages. (It even briefly changed the default settings so that users' posts were visible to "everyone" instead of just "friends"—a nod to Twitter's open model. This, however, brought swift outrage from Facebook users, many of whom valued what Facebook had originally promised—a network of people they knew—and the company backpedaled.)

"As the Twitter challenge emerged, [Facebook] expanded its self-definition further to become a service where people communicate with everyone as well as with their friends," wrote David Kirkpatrick in *The Facebook Effect*. But the dissonance between being a public and a private platform would reverberate for years for several reasons, not least among them the fact that "personal connections packed with very private data may not coexist well with unbridled sharing."

// **CHAPTER 8** //

Tumblr Famous

IT ALL STARTED WITH "FUCK YEAH."

Since its founding in 2007, Tumblr had been a thriving destination for creation and curation, with a scrapbook-like feel. As on Twitter, posts by people you followed were presented in a clear feed—your dashboard—in reverse chronological order. Reblogging—Tumblr's equivalent of the retweet—allowed for memes and posts to go viral across the site, just as happened on other social networks. Co-founder David Karp had decided to hide follower counts as part of the design, so Tumblr didn't tend toward some of the clout-chasing dynamics that emerged on other networks. Instead, content often bounced from blog to blog, iterating as it went. The knowledge that content could spread inspired users to design content with the intention for it to spread as widely as possible. (Among the seeds tossed by Tumblr users were GIFs, or short looping images, and cinemagraphs, a type of animated GIF.)

"The thing that made Tumblr so popular was the reblog feature," said Lina Abascal, a journalist who has covered Tumblr culture. "More people were curating content than creating content on Tumblr. It was a place to craft your own identity out of a curation of others' content rather than crafting your own identity by making your own content."

It was the opposite of the Drudge aggregation model, where users navigate to a single website of curated links. With Tumblr, content came at you from all of the users you trusted. If someone was doing a great job curating great content, that person was worth your attention. This dynamic allowed Tumblr to become the first mainstream social media app where you could gain fame curating other people's content. Seemingly anyone could do so, pseudonymously, with the right subject in mind. Many of the viral Tumblrs that fueled the platform's rise were fueled by one-off submissions and reblogs. And many were devoted to single, funny themes: "Texts from Last Night," "Stuff White People Like," or "Look at This Fucking Hipster."

The most popular craze, however, was a simpler recipe: the words "fuck yeah." "You're on Tumblr and . . . you see something on your dashboard. And you're like 'Fuck yeah! This is what I love.' The term is just so celebratory. I think Tumblr users really identify with it," Amanda Brennan, creator of fuckyeahmodernism.tumblr.com and a former member of Tumblr's community content team, told the *Washington Post*.

Users began to pick any subject they liked: corgis, pizza, Ryan Gosling. Then they'd make a "Fuck Yeah Corgis" site and reblog every piece of content related to corgis they could find. For a lucky few, their sites would go viral after a write up in BuzzFeed or Mashable. And as basic as the recipe was, Fuck Yeah blogs helped transform Tumblr from a niche blogging platform into a viral powerhouse.

The trend arose early in Tumblr's existence, but it didn't blow up right away. "Fuck Yeah" had already emerged as a popular phrase on the internet in the 00s. Its origins lay in the song "America, Fuck Yeah," from the soundtrack to Trey Parker and Matt Stone's 2004 flick *Team America: World Police*, which found new life when a clip of the song was uploaded to YouTube in February 2007. The video racked up hundreds of thousands of views, and in the comments section, people expanded on the meme. "Fuck Yeah," someone would comment, followed by random nouns.

Just a couple months later, in April 2007, the first "Fuck Yeah" Tumblr was created: fuckyeahtechnospeedandgirls.tumblr.com. According to Tumblr, the site went largely unnoticed. But by the end of the year, people had begun to create "Fuck Yeah" Tumblrs devoted to celebrities. In December 2007 a user created fuckyeahsteelydan.tumblr.com, the sixth ever "Fuck Yeah" Tumblr, and the trend slowly grew.

A year later, in October 2008, there was a sea change. Freelance writer Ned Hepburn had built an early niche following on Tumblr and—one fall day, when he was off work—he got high on marijuana and decided to create fuckyeahsharks.tumblr.com. Like all "Fuck Yeah" Tumblrs, he proceeded to reblog anything he could find related to his topic: shark GIFs, shark trivia, shark movie stills. Hepburn would monitor the #shark tag on Tumblr and scan for shark-related posts. He was a curator, netting the best shark content from across the Tumblr sea.

"Fuck Yeah Sharks" took off, and Hepburn gained tens of thousands of followers, a monstrous number at the time. Then the prefix really took hold. "It's a clever and effective way to create a kind of digital shrine to something you're super pumped about, whether it be an animal or an era or the shipping of Harry Styles and Louis Tomlinson," Jessica Bennett, former executive editor of Tumblr Storyboard told the *Washington Post*. "The beauty lies in the simplicity."

Several months later, in April 2009, Tumblr users were creating about twenty-five new "Fuck Yeah" Tumblrs a day. Most "Fuck Yeah" blogs—like most Tumblrs in general—were anonymous. "There were definitely these Tumblr celebrities," said Cates Holderness, head of editorial at Tumblr, "but they were always pseudonymous." For many, posting on the internet was still considered slightly cringy. Some people didn't want to reveal their offline identities because they worried it could interfere with their jobs or school. It was also far easier to get followers when people didn't know who you were, when it was all about Sharks or Steely Dan.

"Fuck Yeah" Tumblrs were usually positive. Their names were easy to remember. Their links were easy to share. All of these were key ingredients for viral success. After a few years, Tumblr hosted hundreds of thousands of "Fuck Yeah" blogs alone, and they became so synonymous with the Tumblr brand itself that the company even named its 2015 SXSW party "The Fuck Yeah Tumblr Party."

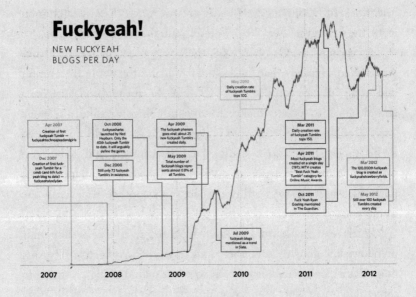

The top "Fuck Yeah" sites were among Tumblr's bigger blogs, and the content creators behind them became minor internet celebrities, with cult followings and real influence around the topics they covered. That might not have amounted to much for, say, sharks. But for the creators of "Fuck Yeah Menswear," it became a real opportunity.

With the platform's emphasis on photos and sharp design, Tumblr had become a hub for street-style blogs, often luring bloggers over from other platforms. In 2010, a stream-of-consciousness fashion blog run by two friends, Lawrence Schlossman and Kevin Burrows, combined two poles of Tumblr: "Fuck Yeah" and fashion. Their blog, "Fuck Yeah Menswear," found a niche by making light of the dominant fashion-blogging schtick while also broadcasting its own sense of style and playing host to street-style mainstays like fashion consultant Nickelson Wooster. In its heyday, a reblog by "Fuck Yeah Menswear" could sell out a coat.

But even as their Tumblr efforts were driving sales and shaping taste, the "Fuck Yeah Menswear" founders weren't able to monetize to the extent that future influencers would. (Schlossman eventually became a wildly successful podcaster, co-hosting a fashion podcast called Throwing Fits and remains a top menswear influencer to this day.) Their

Tumblr came before the Instagram-driven boom of sponsored content, and Tumblr never rolled out an ad network or revenue tools for its creators. It was impossible to see how many followers a given blog had from the outside, which made it difficult to see how influential any one site was.

Success for "Tumblr famous" creators didn't exist on the site itself; it lay in exposure. If a site like "Fuck Yeah Menswear" went viral, its best bet was to head to the gatekeepers of old and cash in their online success for offline prestige. In this era, that often meant parlaying viral posts for job offers at established media companies or their new-media counterparts. Or it meant approaching literary agents and book publishers, in hopes of translating an online readership into a paying book-reading audience.

"Fuck Yeah Menswear" ended up taking this latter route, along with many top Tumblr blogs of the era. There were books created out of not just "Fuck Yeah Menswear" but all manner of memey Tumblrs: "This Is Why You're Fat," "Garfield Minus Garfield," "Hipster Puppies," and perhaps the only lastingly successful one: "Humans of New York." In 2010, there were so many Tumblr book deals that someone created an entire Tumblr dedicated to tracking them. In the same period, the first crop of YouTube stars was also turning to publishing as one of the few sure sources of a payday that could follow from online fame.

※

From 2009 to 2012, Tumblr grew from 6 million monthly users to 34.2 million, putting it among the top 50 websites in America. It also served as a kind of middle ground between two eras: the reign of the blogosphere and the dominance of social media, which was fitting, because Tumblr itself was a hybrid platform. The platform spawned the most popular memes of the era and played host to some of the first truly successful social media creators and influencers. The vast majority of such "Tumblr famous" creators never figured out how to make their fame last or to parlay it into truly successful careers as creators. While they got further than MySpace's "e-celebs," the path of an independent creator or influencer hadn't yet matured.

While Tumblr remained smaller than rivals Facebook and Twitter, it provided a creative environment that neither quite matched. Twitter was so short-form that it could rarely be a creator's main home; they would often need to link to a blog, a YouTube channel, or an old-media column. Facebook's revamp left only two speeds: private posts meant to stay among friends, or public content that competed with everything else on the platform for engagement. If you wanted anything in between, your top option was Tumblr.

✳

Despite impressive growth in the 2010s, the internet's creative hubs—Tumblr, YouTube, and the blogosphere—were still mostly self-contained bubbles. Meanwhile, Facebook and Twitter were trying to become home pages for the internet, collecting the most popular content on the web and feeding it to users. There was a resulting arbitrage opportunity between the creative hubs and the social internet. BuzzFeed or Mashable could dive into those bubbles, harvest the best work of creators, and repackage it for maximum virality on Facebook. These sites developed a symbiotic—though some might say parasitic—relationship with Tumblr and its creators. While a site like BuzzFeed could direct major traffic to Tumblr accounts, ultimately it was BuzzFeed that stood to make money off of it.

BuzzFeed made most of its revenue through native advertising and sponsored posts. It managed to do so on an industrial scale by tapping networks like Tumblr. It would attach, say, Verizon's logo to a post titled "10 Epic Messages That Prove You Should Never Text Your Ex." Then, it would populate the post with GIFs and memes that its editors gathered from around the web, often through social hubs like Tumblr or Reddit. BuzzFeed then repackaged it all using pithy headlines and lists that were optimized—often through real-time A/B testing—to the new algorithms used by Facebook and Twitter. If sponsored posts failed to get sufficient traction on their own, BuzzFeed would simply buy Facebook promotion, ensuring that its brand sponsors got the page views they were paying for.

This business model was particularly suited to the early days, when the entire internet was making its transition to social media, and it worked like a charm—except for the creators who were doing the unpaid work of creating and curating the internet's best content.

While BuzzFeed, Mashable, and other new media aggregators capitalized on this arrangement in the first half of the 2010s, the most quintessential example of the phenomenon came slightly later. In February 2015, a twenty-one-year-old in Scotland named Caitlin McNeill posted a photo of a dress—*the* dress.

Like many on Tumblr at the time, McNeill was a huge fan of popular YouTube stars, so much so that she ran a Tumblr fan page called swiked.tumblr.com, dedicated to a woman named Sarah Weichel, a talent manager who represented many of them. On February 15, McNeill posted about something else: a photo of a black and blue striped lace dress. The caption read: "guys please help me - is this dress white and gold, or blue and black? Me and my friends can't agree and we are freaking the fuck out."

It was an uncanny optical illusion: to some people, the image of the dress looked blue and black; to others, the image appeared white and gold. Sometimes it would look one way, then flip and look the other way, which resulted in endless debate. "The Dress" soon amassed thousands of reblogs on Tumblr as people debated the "real" color of the item of clothing (which was actually blue and black, for what it's worth).

Cates Holderness worked at BuzzFeed, where she monitored Tumblr for viral content. Tumblr had been very good for BuzzFeed, providing a huge amount of content that BuzzFeed turned around and repackaged as its own. (BuzzFeed was so reliant on Tumblr content that it once crashed the company's 2012 Internet Week party and asked people who were popular on the site to take photos in a giant banana hat; the photos promptly turned up on . . . BuzzFeed.) She was wrapping up work for the day when she noticed something pop up in the "Ask" box, which was essentially Tumblr's version of a direct message. The users wanted Holderness to settle an argument, to weigh in on whether the dress in McNeill's post was blue and black or white and gold.

It looked blue and black to her, but she called over her colleagues to take a look. As Holderness watched, they began arguing with one another. Holderness later told *Digiday*, "I realized if we were having this discussion, other people would, too."

She whipped up a post on BuzzFeed titled "What Colors Are This Dress?" Once again, it was simple arbitrage, transferring the social media fervor from one platform to the rest of the web. "There's a lot of debate on Tumblr about this right now, and we need to settle it." The BuzzFeed article reached an unprecedented number of people. As the site repackaged the Dress and sent it out across Twitter and Facebook, it became one of the most viral pieces of content on the internet.

"More than anything it really underlines the differences between Twitter, Tumblr, and Facebook," viral guru Neetzan Zimmerman, who was an expert at creating content that would explode on the internet, told *Vice* at the time. "It went viral on Tumblr yesterday, on Twitter today, and will probably hang around Facebook for the rest of the week. This seems to be the new Viral Cycle. (I'd note that this is at least partly because BuzzFeed gets a lot of its material from Tumblr.)"

Soon, celebrities, including Taylor Swift, were weighing in. The post attracted nearly 30 million page views in under twenty-four hours. BuzzFeed ended up having an officewide Champagne toast to celebrate the record-breaking numbers.

"The Dress changed a lot of things about the notion of virality," Holderness told me. "It was about how people could latch onto these viral moments, whether it was for monetization or whatever. Brands were like, 'OK, wow, this is the impact this one viral thing can have.'"

The irony is that it also signified the end of an era when people relied on digital media for this type of content. As more and more people became regular users of social media, and as those platforms built out algorithms to surface content relevant to each user, viral content no longer needed a BuzzFeed post to reach the masses. Likewise, as more and more creators embraced platforms where they could generate revenue directly, there was less opportunity for aggregators to take a piece of the pie. If the Dress had been posted today, it likely would have gone viral on its own, on the platform where it had originated, without the need for a BuzzFeed-like middleman.

"The tech bubble of the early 2010s really contributed to the rise of influencers or online creators because you got these Mashable and BuzzFeed round ups," said Holderness. "There was a whole economy around content like that."

Caitlin McNeill never got rich off her post, although the company behind the Dress sold out of the item and BuzzFeed celebrated its record-breaking traffic. And while McNeill's Tumblr got lots of new followers, they weren't really interested in Sarah Weichel or YouTubers, which is what her Tumblr went back to covering.

※

In those years, the best thing that could happen as a result of your Tumblr going viral was landing a full-time job, usually in digital media or tech. Sites like BuzzFeed, the Daily Dot, and Mashable made many hires off Tumblr. Early BuzzFeed meet-ups and digital media parties of that era were packed full of "Tumblrebrities" with an eye toward career elevation.

Of course, not everyone who became famous on Tumblr wanted to translate that into a media job, even at one of the new media companies that frequently (in some instances, incessantly) claimed they were nothing like their legacy ancestors.

One insider, James Nord, co-launched one of the first influencer marketing platforms on the back of his experience on Tumblr. Nord created an account in October 2007 when he was fresh out of college and trying to figure out his life. (He was so early that he created a Tumblr blog with his own name, before it was standard practice to do so anonymously.) When Nord joined, the platform was small enough to feel like a community. Posting amateur photography, he took lots of street style photos and city scenes, and he began experimenting with simple effects like filters. Before Instagram existed, this made his photos especially striking and shareable. As he gathered reblogs, the Tumblr team made him one of their recommended accounts, and by 2011, Nord had become one of the most followed photographers on the platform.

Nord had a marketing job for a stock photo company as a day job, and one of his theories was that the online photography community was completely undervalued. He tried cold-calling brands to make a pitch for featuring products on his page now that he had thousands of followers—but no one was willing to bite. Nord caught a break when Tumblr recommended him and several other bloggers for a shoot at New York Fashion Week.

Fashion was one of the first industries to recognize the usefulness of bloggers and social media power users, and Nord became one of the first Tumblr users who was welcomed in. But it was still mostly unpaid work. True to the era, Nord's compensation was the exposure from working with fashion icons.

Nord couldn't ignore what other early bloggers had come to grasp: they *could* drive sales, if brands would just give them the chance. The issue on Tumblr was that it was impossible to keep track of any sales. Brands couldn't tell how big certain fashion blogs were, since audience numbers were hidden.

Tired of waiting, Nord and a few partners started a business in 2012 that would connect brands with fashion photographers. The first iteration was called Fohr Card, with the purpose to simply provide brands with a reliable list of bloggers who'd shoot products, along with the back-end numbers of the bloggers' audiences and engagement. From there, Nord's company evolved into a full-fledged influencer marketing platform, now called Fohr.

He did so just in time, because Instagram was putting the finishing touches on the internet-wide transformation that Tumblr had started. Tumblr was going to be left behind.

Despite launching as a mobile app in 2009, Tumblr never really innovated on its product or made a significant design overhaul as the rest of the world moved to mobile. Even in 2023, the platform looks almost exactly the same as it did over a decade ago. It's still a place rich with meme culture, but with fewer and fewer trends reaching the mainstream.

In 2013, Tumblr was sold to Yahoo! for $1.1 billion, a highpoint that coincided with the beginning of the company's decline. What had been a cool, culturally relevant, New York start-up was suddenly owned by a decaying Web 1.0 giant that had no clue how to run it. Many of Tumblr's most influential employees left the company. Many Tumblr users also moved on, this time to the next hot property that offered not just a new visual medium but a whole new online landscape.

// CHAPTER 9 //

Instagram's Influence

ONE NIGHT IN JANUARY 2011, LIZ ESWEIN, A TWENTY-TWO-YEAR-OLD senior at New York University, was sitting at home scrolling through Twitter. Eswein had spent a lot of time online during her senior year, after a bout with Lyme disease left her with autoimmune issues. She was worried about what type of job she'd be able to hold once she graduated that spring.

A tweet caught her attention: it included a link to an article about a new photo-sharing app called Instagram, which was raising money from investors. Instagram had launched barely three months earlier as one of the first social networks that was mobile-only. It had already grown to nearly 2 million users, who were collectively uploading over a quarter million photos each day.

Eswein had an iPhone—Instagram was iOS-only at the time—so she decided to give it a try. She attempted to register a few different handles on the service, but each came up as unavailable. "I thought to myself, I live in New York City," Eswein said. "So, I typed in '@newyork city' and it registered. I didn't think anything of it, and I just started posting." Her first photo, uploaded on January 11, 2011, was a picture of a plate of bacon and eggs, made vibrant by one of the app's early filters.

Instagram was the brainchild of Kevin Systrom and Mike Krieger, both recent graduates of Stanford. It wasn't their first start-up idea. They were originally working on a more complicated, Foursquare-like app called Burbn, but a series of meetings with VCs prompted Systrom to pivot and simplify the product.

Three years after its debut, the iPhone was the hottest piece of technology around, and it wasn't because it made phone calls better than its rivals. Consumers prioritized camera quality when shopping for smartphones. Facebook and Twitter had established themselves as the hubs of social media. Why not make a bet that the two trends would combine—that the phones in everyone's pockets would become the lenses through which we saw the world?

There were plenty of photo apps, with a number of social networks trying to elbow into the growing market. But none had managed to combine photo editing and social networking well. Wireless networks were still slow, and the iPhone camera took grainy photographs. To make a photo-sharing app work, the Instagram founders had to be tactful. Upon launch, they imposed a Twitter-like constraint, limiting photos to a 306-pixel square to limit data costs. They also created a range of filters, so that everyone's photos could appear bright and distinctive.

While developing Instagram in 2010, Systrom and Krieger also made sure to keep the app easy to understand and flexible, including the capability to link to users' Facebook, Twitter, or Tumblr accounts. As a result, it looked like a complement to existing social media sites, rather than a competitor.

By the time the platform launched, it had attracted influential users like Jack Dorsey and others who were already becoming accustomed to sharing the minutiae of their lives online. Dorsey and others posted their Instagram pictures to Twitter and marveled at how good they looked, and the app took off. In fact, it grew faster than any other social media site, attracting 25,000 users on its first day, then 100,000 users within the first week, 1 million within the first two months, and 2 million less than a month after that.

Snoop Dogg joined in January 2011, becoming the first major celebrity on the platform. Already an avid Twitter user, Snoop was one of many A-listers who were drawn to the directness and ease of using social media to connect with fans. Instagram was easier to use than Twitter, especially for celebrities, who were often wary about sharing text. There was no need to think of a witty thing to say in 140 characters. On Instagram, you could just snap a picture, apply a filter, and post it—no words necessary.

Snoop Dogg's first post hinted at the conflict to come. To his 2.5 million followers on Instagram, he shared an image of himself holding a can of a new drink—Colt 45's 4Loko competitor, Blast—with the caption, "Bossin up wit dat Blast." The first post from a major celebrity was also, as Sarah Frier wrote in *No Filter*, "the first case of an ambiguous advertisement on Instagram."

This conflict was practically built into Instagram: if millions of people would be staring at a wall of beautiful images, some would be tempted to turn that wall into a billboard. From Instagram's inception, Systrom had been unambiguous about where he stood on this temptation: strongly against it. In part, he was following the Silicon Valley playbook of focusing first on growth. Companies like Facebook, Twitter, and YouTube considered monetization only after they achieved huge scale. But Systrom's thinking was also aesthetic. A photographer himself, he understood that Instagram's early appeal lay in allowing its users to see their world anew.

It was this feeling that sparked Liz Eswein's interest. In her first year, she posted sporadically about her life as a twentysomething in the city: yellow cabs, the Empire State Building at night, and snowy downtown streetscapes. She also began to connect with other Instagram users, attending local meet-ups and spreading the word about the app.

These early meet-ups were a key part of the app's strategy. Systrom and his small team took advice from one of the company's investors, Steve Anderson, to heart: "Anyone can build Instagram the app, but not everyone can build Instagram the community." So, when early Instagrammers began to organize #Instameets or #Instawalks to connect with others on the app, the company supported these efforts, publicizing meet-ups in cities around the world and the superfans who attended them.

With this in mind, Instagram's first employee and head of community, Josh Riedel, reached out to Eswein and invited her to a Google group with other Instagram power users. The group included Philippe Gonzalez, a Frenchman living in Madrid who had started @Igers and a blog at Instagramers.com, both of which highlighted content creators who were making the most of Instagram. There was also Laurie Keiko, an Instagram power user in the East Bay who hosted monthly photo walks. To Instagram's founders, users like Keiko and Eswein represented the ideal that Instagram was looking for: photographers who had enjoyed practicing their craft on the app and wanted to share that joy with others.

As Instagram grew through 2011, it faced fierce competition. Advertisers were angling for ways to capitalize on the app, but Systrom was not welcoming to them. And celebrities, who were used to getting paid up to seven figures for tabloid pictures, wanted compensation if they were going to post.

Instagram's biggest account after six months in existence was Justin Bieber. The young pop star, who had been discovered on YouTube, was so influential that whenever he posted, the shockwave of traffic from his fans threatened to crash the whole site. Realizing his power, Bieber's manager, Scooter Braun, gave Systrom an ultimatum: Let Bieber invest, pay him for his content, or else he'll stop using Instagram.

Systrom knew that having stars on the app would fuel growth, but he didn't want to pay users or overcommercialize the site. He preferred to keep the app open and fun, so he called Braun's bluff. Bieber briefly left the app but returned to the platform after a short time. The pop star was pulled back by the all-consuming drama in his relationship with actress Selena Gomez. Before long he was posting again and his follower base was bigger than ever. At one point, Instagram was forced to expand its infrastructure simply to handle the flood of engagement that Bieber's posts alone would generate.

The Bieber kerfuffle was a symptom of Instagram's runaway growth. Much like Twitter, the app was becoming a hub for pop culture, but since Instagram was acquiring users at double the speed, it was in a unique position to consider monetizing sooner than others. The founders were confident they could generate cash by embracing ads. But they knew that taking the wrong route could easily destroy the site's appeal.

While Facebook and Twitter were building their platforms around virality, Systrom and his team instead decided to favor aesthetics over growth. This strategy worked. Instagram's user growth accelerated, and as the Bieber gamble had shown, celebrities simply couldn't afford to not use the app. Facebook was flummoxed as to how to compete. While Facebook didn't have a strategic response plan, it did have a lot of money. The company figured if you can't beat 'em, buy 'em. So, in 2012, Facebook bought Instagram for $1 billion.

Post-acquisition, Instagram fought to chart its own path and remain as independent as possible within Facebook's growth-obsessed culture. Systrom rejected repeated calls to add a Reshare button, a proven way to increase engagement by allowing viral content to race across the platform and boost users' time on the app. He insisted that it would gum up the feed and prevent each user from feeling like their own personal curator.

Likewise, Instagram refused to run ads for as long as it could manage under the Facebook umbrella, holding out until November 2013. Even then, Systrom helped orchestrate the first slate of advertising so that it aligned with his vision of Instagram's style. He fought to keep the ad network distinct from Facebook's cheaper fare. Furthermore, while Facebook and Twitter had embraced a neutral, algorithmic approach to content, letting whatever was popular rise to the top, Instagram chose to be more selective. The app had its own "Popular" page, featuring the most trafficked posts on the app, but the company avoided promoting it. Early on, it became clear that the Popular page surfaced low-quality content ("boobs and dogs," as chef influencer Jamie Oliver put it). That wasn't what Instagram's community team wanted to promote.

In response to these concerns, the community team tried a different approach. They would handpick accounts that they found interesting, highlighting users that matched the aesthetic ideal of what they thought Instagram should look like. One of the first such sifting methods was to curate lists of "suggested users," many of which drew from

Riedel's power-users Google group. In 2012, the company began using its own account, @instagram, to highlight inspiring trends, images, and accounts. Such promotion became a key lever for driving popularity on the site—one that was all the more powerful in an ecosystem without resharing. The taste-making @instagram account soon became the biggest account on the site.

All the while, Instagram was aggressively recruiting celebrities to join the app. It hoped to combine star power with high-quality user-generated content to give the site an elevated, aspirational image. For a while, star power kept the site in the news, while featured users showed everyone else how to use the app to the fullest. Instagram did become an aspirational place, not just full of celebs but all sorts of interesting people.

But not everything went as planned.

The downside to promoting a polished aesthetic was that Instagram undermined the casual, stress-free posting that had made it so fun at the start. What's more, Instagram's approach *didn't* prevent the site from filling with ads. Instead, it simply changed who brought those ads to the site.

Ultimately, by banning ads while curating its amateur userbase, Instagram incentivized brands to find a side door. Instagram helped hundreds of curated users like @newyorkcity gain hundreds of thousands of followers, and those users saw what others online, like bloggers and YouTubers, could do with an audience and online advertising. By supporting the Liz Esweins of the app, Instagram had inadvertently created an influencer factory. Now it didn't know what to do with it, and a bunch of those influencers were done waiting.

※

Fashion brands and consumer goods companies were among the first to embrace Instagram, which really did make *the* perfect marketing venue. The platform was image-based, aspirational, social, and full of young shoppers. Brands created their own accounts on the site and started interacting with users. In 2011, the biggest branded Instagram account was Starbucks, with around 117,000 followers and over 57,000 photos tagged with #starbucks on the app.

But there were still no ads for sale, no reshares, no paid promotion tools. While companies could make their own accounts, most people weren't coming to Instagram to follow a bunch of brands.

Certain companies figured out clever tricks, though, many of which involved working with Instagram's homegrown talent. Red Bull developed a follower base on Instagram in the same way that an average account would: it posted images that were striking on their own rather than advertisements for products. As the account approached 100,000 followers in 2011, it took a page from Twitter—specifically, from Ashton Kutcher.

Red Bull challenged @newyorkcity in a race to 100,000 followers. Eswein won the race, notching another victory for the underdog. As a prize, one of her images was posted to the Red Bull account. When her black-and-white street scene from New York City appeared on @redbull, both profiles helped each other spread their reach. In essence, Red Bull had placed an ad on @newyorkcity by drawing the attention of Eswein's 100,000 followers.

Brands quickly created their own workarounds to bypass the barriers that Systrom had created. Brands began reaching out to the accounts Instagram had featured as suggested users and started sponsoring individual posts, which blurred the usual lines between average users and professional advertisers—and inadvertently encouraged its users to *become* advertisers.

※

Over the course of 2011 and 2012, a new class of Instagram-native creators joined the celebrities who had already leaned into sponsored content. Eswein knew that bloggers made money through advertising and endorsements, and as her following grew, she began to explore those opportunities as well.

She also encouraged her family members to join the app. Her mother and brother signed on early enough to nab the handles @Food, @Baking, @realestate, and more. "My mom was in food and my brother was in real estate," Eswein explained. They nabbed other handles too. Together, they would become, as the *New York Times* heralded in 2014, the "First Family of Instagram."

Eswein was the first of her family members to try to monetize her following. In 2012, when she reached 200,000 followers, she realized that she had a larger following than many magazines, whose pages carried a full slate of advertisements. She could reach more people, she knew, so she began to pursue brand deals directly. One of her first deals was a Nike campaign, where she made less than $100 to post an image on her feed. She quickly realized that she could do better.

Eswein and two other top Instagram creators, Anthony Danielle and Brian DiFeo, started a small social media marketing agency called the Mobile Media Lab, in order to cater to brands interested in Instagram. They connected companies engaged in early social media marketing (among them Samsung, *Lucky* magazine, and T-Mobile) with the founding creators as well as other influencers they'd come to know, forming partnerships.

Meanwhile, Eswein's brother trekked around Manhattan, meeting with restaurants and food brands to espouse the app's potential for businesses. Paying creators to post advertisements on Instagram, he argued, was like posting on a digital billboard—the very concept that Systrom had wanted to avoid. By 2014, Eswein's salesmanship resulted in partnerships with numerous restaurants and Fairway, a local supermarket chain.

Over time, Eswein and her peers helped create a new model for advertising campaigns, one that relied on creators to handle all production themselves. For instance, when Nordstrom rolled out its new line of fall handbags and accessories in 2014, the company hired a group of Instagrammers to photograph the collection and write product testimonials. Eswein was featured prominently among eleven other Instagrammers with large audiences.

By 2014, Mobile Media Lab had expanded to work with top Instagram influencers, booking campaigns with hotel chains and fashion brands. But even with their growing client base, Eswein was far from alone: 2010 to 2014 was a golden era of growth on Instagram. Influencers raked in millions of followers. Much of that growth was driven by the app's suggested user list. Many of those suggested users would follow Eswein's lead and turn Instagram into a full-time job, centered on sponsored-content campaigns, all before Instagram—technically—allowed advertising at all.

※

The company wasn't blind to what was happening. They didn't love that the creators they promoted were monetizing their followers' attention, especially since they were meant to be models of how to use the app. In the summer of 2012, the platform made a first cut to its "suggested user" list, removing almost two-thirds of the two hundred accounts it had been promoting. "Instagram was not supposed to be about obvious self-promotion," Systrom said. Instead, he argued, it was supposed to look elevated and authentic.

But now that Instagram was owned by Facebook, no one at the company believed they could avoid running ads forever. Nor did they think that the app could survive long-term without welcoming brands on some level. At this point, they just wanted to keep the commercialization subtle.

Once they opened the gates, the bloggers came rushing in.

By the early 2010s, the blogosphere had matured beyond banner ads, adding a new set of tools to allow a middle class of bloggers to thrive. Blogs had also grown increasingly visual, especially among lifestyle, fashion, and mommy blogs. As Instagram established itself, the bloggers invaded, bringing their monetization schemes with them.

The root of its transformation was an innovation that harkened back to Dallas, Texas, and to 2010, where a fledgling blogger was trying to get her bearings.

Amber Venz (now Amber Venz Box) had wanted a career in fashion since she was a child in Dallas, Texas. She was a go-getter, starting small businesses throughout her youth and eventually launching a local jewelry line while attending college at Southern Methodist University.

As an undergrad, Amber applied to intern at the glossy fashion magazines she idolized in New York City. But she couldn't get a callback. It was the recession, and only a handful of jobs were available. Someone from out of town without connections? She didn't stand a chance.

After graduating, Venz moved in with her parents and got a job in Dallas as a personal shopper affiliated with local stores. She traveled between boutiques and the homes of wealthy Dallas residents, bringing stylish outfits and pieces for them to try on. Whenever her clients made a purchase, she received a commission.

It was decent business, but she still ran her jewelry line. After a few months, she decided to start a blog to promote her products. Like many early bloggers, Venz loved having a creative outlet, sharing her fashion sensibility with readers, and connecting with others who shared her interests, not just in Dallas but around the world.

There was only one problem. Even by 2010, there wasn't much money in blogging, especially for anyone just starting out. Display ads only brought in meaningful revenue to larger websites. Sponsored content was likewise only available and lucrative to those at the very top of the blogosphere. For everyone else, blogging was a tough business.

Venz's boyfriend and later husband, Baxter Box, was running into similar problems. He had started a website devoted to surfing, and had even wooed sponsors like Billabong. But the only compensation he received was free gear, which wasn't enough to turn even a high-quality site into a business.

While Venz tried to figure out how to turn her passion into a livelihood, her blog was cannibalizing her day job. Venz liked to post about her fashion finds and favorite retail deals. Her clients learned that they could simply browse her website rather than schedule an in-person appointment. Many clients thought they were supporting Venz by purchasing the items they found listed on her site, but they were in fact cutting out the personal-shopping commission that Venz relied on.

When Venz caught onto this, she came up with an idea. Why *shouldn't* she receive a commission on the sales she drove online? Personal shoppers were tastemakers who got paid for their style; fashion bloggers were essentially doing the same, for more people, but for free. What if she could change that?

This breakthrough led to the launch of RewardStyle in 2011. Venz and Box teamed up to create a company to use affiliate marketing for bloggers. RewardStyle built partnerships between online retailers and bloggers: when bloggers posted about a given outfit or style, they would use unique links to the retailers' online stores. The service was a win-win, allowing retailers to learn where their traffic was coming from, and when a blogger like Venz linked to an outfit that her readers purchased, she would get a cut of the sales.

It didn't take long for Venz and Box to set the whole thing up; Box was an engineer by trade and knew some coders, and Venz, the fashion blogger expert. Upon launching the company in 2011, they started inviting influential bloggers, sparking a significant increase in referral traffic and sales to online retailers. Now, there were metrics and dollar signs showing just how big online tastemakers had become.

RewardStyle allowed more bloggers to make a living online. And because the service was an invite-only network, it was reliable and trustworthy. It was so successful that it didn't take long for rival services to pop up, which meant even more revenue started making its way into blogger bank accounts.

Thanks in large part to Venz and Box, affiliate marketing became a mainstay of the online creator industry. Private blogging networks, which hosted certain "top tier" bloggers who had a large audience and commercial marketability, also emerged, acting as agents and negotiating advertising contracts on behalf of bloggers. They also provided PR services, helping bloggers score invitations to fashion weeks and exclusive industry events.

Two years after RewardStyle had launched, several bloggers were making nearly a million dollars a year from the company's programs. RewardStyle had become an influencer marketing behemoth.

Always on patrol for the next thing, Venz noticed that with Instagram on the scene, some RewardStyle members had cut back on blogging or were skipping it entirely. Display ads and click rates on blogs had been in decline for years, and bloggers, realizing that the ad business had shifted to mobile, raced to Instagram to get in on the boom. Influencers could now exist solely on Instagram and drive mammoth numbers.

While early Instagrammers considered it gauche to post professional-looking photos, bloggers had already embraced the photoshopped look and promptly took it to Instagram. The bloggers' polished aesthetic fit well with Instagram staff's preferences. Top lifestyle bloggers were already staging high-quality photoshoots for their blogs, so transitioning to Instagram didn't require much. But while the lifestyle bloggers adapted well to Instagram, their business models—affiliate links and content campaigns—were largely absent from it. Recognizing the gap, Venz and RewardStyle developed LIKEtoKNOW.it (LTK). LTK offered a custom-built affiliate program for Instagram, allowing users to link to any product in a given Instagram post. The bloggers' affiliate links could now move over to Instagram, and bloggers could then leave the blog behind, or at the very least expand tremendously beyond it.

Many bloggers were willing to make the jump. They found Instagram (and other social apps like Pinterest) easier and more streamlined than blogging: they didn't have to write as much, and every post automatically reached a built-in audience. Their original blog remained as a vestigial home page, if it remained at all.

LIKEtoKNOW.it became a critical piece of the Instagram influencer economy. Like RewardStyle before it, LTK allowed even small or medium-sized accounts to develop real revenue streams and commit to Instagram full-time. The cash flow from their LTK transaction revenue could lead to campaigns with major brands, with LTK's metrics proving to potential partners that an Instagrammer could drive meaningful sales.

Those metrics reaffirmed that online influencer campaigns were more effective at moving the sales needle than just about every other form of marketing. Social media users adored the lifestyle creator accounts they followed. If a beloved account integrated a product—especially in a glamorous photoshoot in some far-flung locale—followers would leap at the opportunity to buy it up.

✳

This infrastructure snapped into place before Instagram ran its first official ad, which happened on November 1, 2013. Systrom personally oversaw the launch, a post from luxury brand Michael Kors. Yet even after jumping into the promotional pond, the platform continued to drag its feet when it came to supporting the sophisticated monetization tools that its rivals offered.

Engagement on Instagram had been tricky to measure in its early years. The app didn't offer robust analytics for personal accounts, so most sponsored-content campaigns initially ran based on follower numbers. If you had 100,000 followers, you could charge a decent rate even if your posts themselves didn't get tons of likes. The launch of LTK and a slew of influencer marketing platforms helped change that. At the same time, users with a sharp eye began to find ways to expand their reach. Instagram's feed was in reverse-chronological order, so content creators didn't have to worry about catering to some algorithm.

"It was really easy to show up in people's feeds," Sarah Frier noted in *No Filter*. "If you were a creator you just had to post at certain times of day. That allowed a lot of people to see amazing growth."

Many Instagram stars would engage in "follow for follow" campaigns (where if you followed a user, they'd agree to follow you back) or fill their captions with hashtags in order to boost their numbers and generate more engagement. Celebrities simply bought legions of followers from spammers.

Instagram's December 2014 crackdown on spam and fake followers became known as the "Instagram rapture." In the mass purge, celebrities like Kendall and Kylie Jenner lost hundreds of thousands of followers;

others lost millions. Akon's follower count dropped from 4.3 million to 1.9 million. Rapper Mase lost so many followers he deleted his account entirely.

But content creators, unlike traditional celebrities, emerged from the purge largely unscathed. However, Instagram put out a statement warning that spammy behavior or excessive self-promotion could get a user banned. "When you engage in self-promotional behavior of any kind on Instagram, it makes people who have shared that moment with you feel sad inside," the company said. "We ask that you keep your interactions on Instagram meaningful and genuine."

Instagram's early product principles initiated a transformation of the internet as a whole. Creators, advertisers, and everyday users had turned away from the disjointed web of the 2000s and flocked to social media. Instagram proved not just that the future was social but that it was also mobile and visual. And that it *wasn't* the "friends over fame" vision that Facebook, Friendster, and even Twitter initially espoused. Instagram cemented the lessons that Twitter had haltingly learned: users were drawn to social networks that went *beyond* friends and featured celebrities, pop culture fixtures, niche authorities, and dramatic personalities.

Perhaps the greatest irony in all this was that the people who opened the door were not those who walked the red carpet. Instead, just as was the case with every platform that came before it, Instagram was saved by the rule breakers. *Because* Systrom had meticulously avoided making Instagram a billboard from day one, it now was a venue where self-promotion and stealth ads were the dominant currency.

In the decade to come, social media would become intertwined with fame and entertainment, with the top platforms battling to make sure that people (famous and not) went to their platform first.

Twitter had the head start. At the 2014 Oscars, host Ellen DeGeneres took time out of the main show to gather a crowd of top movie stars around her and take the mother of all selfies, including Brad Pitt,

Meryl Streep, Julia Roberts, Bradley Cooper, Lupita Nyong'o, and Jennifer Lawrence among others. The tweet immediately became the most-retweeted in the app's history.

That fall, Instagram fired back in its own play for cultural capital. It dominated the cover of the September issue of *Vogue*, the fashion world's most coveted real estate, with a headline that read "THE INSTAGIRLS!" The featured models—Joan Smalls, Karlie Kloss, Cara Delevingne, and more—were not just stars; they were talking enthusiastically about how Instagram had helped *make* them stars.

PART IV

The Platform Battles for Creators

// **CHAPTER 10** //

Vine Time

OF THE VIDEO APPS THAT SPRANG UP IN THE MID-2010S, VINE WAS the most important. It saw the fastest growth, established itself as a cultural tastemaker, and cultivated stars who toured the country to massive audiences. Vine was TikTok before TikTok. In 2014, Vine owned the market despite challenges from multiple rivals. The company's only problem was itself.

Rudy Mancuso grew up in Glen Ridge, New Jersey. He lived with his parents and younger sister in a small home on a cul-de-sac on the edge of town. His Italian American father owned a beauty salon in Montclair, the next town over. When Mancuso wasn't in high school, he spent most of his time at home with his Brazilian mother. Mancuso spoke fluent Portuguese and Italian from a young age, but his learning disabilities caused him to struggle in school. He was diagnosed with synesthesia, a neurological condition where one sense overlaps in ways with another. The condition could be maddening when trying to concentrate. Still, he pushed through, graduating from high school in 2010 and enrolling at Rutgers in the spring of 2011.

In college, Mancuso decided to study video production and business. At the time, he wasn't very online. He had an Instagram account where he shared photos of his friends and parents, but had no ambitions for it beyond that. Then, in the spring of 2013, a friend told him about a new mobile app called Vine. On Vine users posted six-second video clips from their phone. Intrigued, Mancuso began crafting short skits, and soon he roped in his family members as characters. "I started inventing these silly characters that satirized a lot of people and situations personal to me," he said. After he'd made a dozen videos, Mancuso realized that it wasn't just friends watching him on Vine. He was still interested in pursuing film and theatre, but he began to realize that his dream might take a different form than he'd planned. His numbers were growing exponentially, and his videos were being seen by thousands.

※

Despite its short life, Vine had an outsize influence on our lives today. Founded in June 2012 in New York by three tech entrepreneurs, Dom Hofmann, Rus Yusupov, and Colin Kroll, Vine jumped at an opportunity created by YouTube. The year before, YouTube introduced a basic mobile app. But despite advances in phone cameras, the app did not offer a way to record, edit, and upload video from your phone.

The signs were clear that internet users were hungry for more visual content. Instagram became a billion-dollar giant by making smartphone photos stunning and shareable. Tumblr, a heavily visual platform, would soon sell for over $1 billion. Mobile video appeared to be the next big thing.

Vine's app allowed amateurs to create their own mini masterpieces with easy-to-use, mobile-first editing tools. It instituted a six-second cap on video length for the same reason Twitter and Instagram imposed their own constraints early on: data limits. While this limit seemed pathetically brief to some, the Vine founders noted the ubiquity of GIFs, whose looping animations rarely lasted more than a few seconds.

Vine's promise was so enticing that in October 2012, several months before the app's planned launch, Twitter bought it. The deal reportedly valued the unlaunched app at $30 million. It made sense. Vine's short video posts were a natural complement to Twitter's short text and image posts. Twitter, for its part, had been looking to expand its own video offerings. What was a Vine clip if not a video Tweet?

Despite new ownership, the Vine team stayed heads-down, working nights and weekends in the lead-up to launch.

In beta testing, the Vine team noticed surprising behavior from its users. "Our original beta had something like 10 or 15 people on it, and even with that small group we started to see experimentation pretty early on," co-founder Dom Hofmann told the *Verge*. The six-second constraint prompted users not just to record and post about their day, but also to craft small narratives. "Watching the community and the tool push on each other was exciting and unreal," Hofmann continued. "Almost immediately it became clear that Vine's culture was going to shift towards creativity and experimentation."

On January 23, 2013, Dick Costolo, Twitter's then-CEO, tweeted, "Steak tartare in six seconds" along with a short Vine video of a waiter preparing steak tartare at Les Halles. The steak tartare video was created by Hofmann, who also tweeted it. When *Time* magazine later asked Hofmann why he used a steak tartare clip to demonstrate his new app, he said there wasn't any special meaning in it: "I was on a steak tartare kick at that time." The next day, Vine officially opened to the public.

The app looked similar to Instagram at the time. Users could follow each other and scroll through a chronological feed of videos posted by people they followed. The app's clips—known as "Vines"—spread through Twitter, which featured Vine as its go-to tool for posting video.

Vines quickly began racking up attention. Within just a few months, the platform began to cement itself in popular culture. Its prominence came not from the steak tartare or New York subway clips, but from an emerging cadre of creative young Millennial users.

Jérôme Jarre was one of the first Vine stars. Jarre dropped out of business school in France at age nineteen and moved to China to launch tech start-ups. After some success, he used his earnings to move to Toronto and co-found a software company called Atendy. When he heard about Vine, he downloaded the app and began posting.

In April 2013, Jarre generated the first viral video on Vine. Captioned "Don't be afraid of love," the video opens with Jarre walking through the aisles of IKEA, speaking softly and up close to the camera. "Why is everybody afraid of love?" he wonders aloud. He then approaches a random elderly woman in front of light fixtures and shouts "LOVE!" She jumps away in terror and shouts, "Oh!"

Jarre's Vine perfectly encapsulated what the six-second medium allowed for: a quick set-up, a moment of surprise, and a lasting laugh. The "Why is everybody afraid of love?" Vine was actually based on an early viral YouTube video posted in April 2008 by the channel adarkenedroom featuring a man jumping in front of pedestrians and shouting, "Technology!" Jarre's Vine began spreading across the network and didn't stop. The joke's success soon eclipsed everything else on the platform. Vine had not yet displayed "loop counts," its version of view counts, but Jarre's video gained more than a quarter of a million likes and became an overnight meme.

The video also made Jarre one of the app's first stars. He went from 20,000 followers to over a million by the fall of 2013 as he posted man-on-the-street pranks with clever commentary. In an interview with Ellen DeGeneres that year, he explained his approach: "I've always been doing stuff to strangers in the street. . . . I never recorded it because I didn't have the skills to do YouTube videos, so Vine was the perfect way for me to show it in six seconds and share it on the internet."

By the summer of 2013, Jarre had stepped away from his start-up and was spending twelve hours a day on Vine. As his follower count rose, he connected with other up-and-coming Viners. Soon, Jarre was collaborating with Nicholas Megalis, a singer-songwriter. He collaborated with Marcus Johns, a student at Florida State University who posted comedy videos similar to Jarre's. One of Johns's popular videos, "My dad is such a loser," features Johns attempting a kick flip on his skateboard and failing. His father then walks down the driveway and says, "Hey, I used to ride skateboard." Johns looks at him and says, "My dad is such. . . ." before the camera flips around to reveal his father doing a handstand on the skateboard.

With his multilayered skits and deadpan humor, Johns's following outpaced Jarre's. On July 3, 2013, Johns became the first Vine user to hit 1 million followers. He commemorated the moment with a Vine where he rips off his shirt to display "1M" written in green paint on his chest before tripping and falling to the pavement.

There was also Rudy Mancuso. His skits had begun to attract large audiences, and he'd struck up friendships with Jarre, Johns, and others through group chats. They shared filming tips and together navigated the whirlwind they were caught up in.

Vine's video tools, while usable, were limited. "In the beginning of Vine you had to press the Record button to film and release the finger to stop," Mancuso explained. "There was no editing, there was no putting music in in post-production. What you recorded was what you got." Mancuso made the most of what he had. His Vines were among the first to feature multi-part stories with recurring characters and a "cast" primarily made up of family members. Other Viners admired his film-making prowess and directing techniques.

"The most talented comedian Viner, not at the moment but period, is Rudy Mancuso," Jarre said at the time. "He's also the first Viner I ever saw with a tripod."

In the summer of 2013, kids flocked to Manhattan for Vine meet-ups, where they would get together to create Vines with other young people. The meet-ups were small at first, a few dozen kids here and there. But soon they ballooned to hundreds, as fans of popular Viners tried to catch a real-life glimpse of the figures they saw daily on their phones.

Mancuso, Jarre, and others watched these meet-ups emerge, and after Johns hit 1 million followers, they decided it would be fun to do a meet-up of their own. "We decided to all meet in NYC because most creators were East Coasters. We were talking and realized one day: Holy shit, we're all the top Viners," said Mancuso.

On July 25, 2013, Jérôme Jarre, Rudy Mancuso, Marcus Johns, and Nicholas Megalis put out a flier on Instagram created by another Viner Chris Melberger: "Central Park (Sheep Meadow), NYC on Saturday, the 27th. 3pm. wear a costume #costumeparty." They posted a similar announcement on Vine.

"We wanted to know if we all came together, how many people would come see us and meet us like a true meet-and-greet," said Mancuso. Even though they were some of the most followed people on the platform, the numbers didn't seem real.

The morning of the meet-up, Mancuso made his way to lower Manhattan, where he gathered with the other Viners before heading to Central Park. While on the way, he tried to guess how many people might show up at the event. He figured that of his 100,000 followers, a couple hundred might be in NYC, and of them, maybe a dozen or two would show up.

When they reached their destination, they gripped each other's hands and walked into Central Park. The large meadow in the heart of the park was blanketed with hundreds of people. It took a moment to realize that this crowd wasn't just out enjoying a beautiful summer day. These people were here to see them.

Once the crowd realized the Viners had arrived, it swarmed them. Needing some form of crowd control, they switched locations, walking to another area of the park with a large boulder that functioned as a makeshift stage above the fray. They posted Vines updating fans with their new location. "We said, 'Wherever you are, come here,'" Mancuso remembered. "Seconds later it felt like everyone surrounded the rock."

The fans were nearly all teenagers and children, the demographic most enamored with Vine. They screamed and shouted. Some people cried just at the sight of Johns, Jarre, Megalis, and Mancuso. It was overwhelming. Mancuso and the others posed with fans and snapped photos for Instagram. They crowd-surfed, jumping from the boulder into the throng as their fans reached out and caught them.

"That was when we realized, 'Oh, these people are real!'" said Mancuso.

// **CHAPTER 11** //

A Tangle of Competitors, A New Era for Users

IN A MATTER OF MONTHS, VINE ESTABLISHED ITSELF AS THE HOTTEST social platform of 2013, minting a slew of new web stars. That fall, an eighteen-year-old Ohio University freshman named Logan Paul appeared on the *Today* show and took over its Vine account. Paul had amassed 1.5 million followers on Vine prior to his cameo on *Today*, but the appearance catapulted him—and the app—to a new level of mainstream attention. A month later, Jarre appeared on Ellen DeGeneres's show and spoke about the power of Vine to millions of viewers.

While Jarre, Mancuso, and Paul accrued legions of fans, they were barely on Vine executives' radar. The Vine founders noted the stars' large followings, but they were preoccupied by more pressing issues.

Within Twitter, Vine was run as an independent team within the larger company. Hofmann, Yusupov, and Kroll reviewed every decision at Vine throughout 2013. They wouldn't hire a community manager—their first non-technical employee—until nine months after launch, when millions of people were already using Vine. The result was a company struggling to manage the litany of strains caused by rapid growth while in the glare of the media spotlight. It lagged in rolling out features and the internal structure was chaotic.

Above all, however, the app had competition. Vine was the first app to turn mobile video into a breakout success, but once it proved the concept, mobile video became *the* battlefield of social media.

A few months before Vine's launch, Instagram's Kevin Systrom told the *Verge* that mobile video wasn't ready to take off, due to "a combination of data speed limitations and the time it takes to watch a video. Videos are a very difficult medium to be good at, and also a difficult medium to consume quickly." Vine's six-second maximum addressed these concerns, while better smartphone cameras and faster mobile networks paved the way for mobile video production and distribution.

"I think there's a big opportunity here," said Michael Seibel, cofounder and CEO of Socialcam, another mobile-first social video app, in a 2012 interview with TechCrunch. "Everyone recognizes you're walking around with video cameras 24/7 on you, but people still feel like these video cameras are intimidating and hard to use." Seibel hoped to be the disruptor to win mobile video. Socialcam, which launched in 2011, billed itself as "Instagram for Video" and accrued over 3 million downloads in its first year. Meanwhile, Vine reported over 13 million downloads within its first six months—almost ten times Socialcam's growth rate, but still far from mainstream.

Dozens of promising apps emerged, many with millions of venture capital dollars behind them. Socialcam, Viddy, Telly, Mobli, Klip, Color—these apps battled one another and Vine for supremacy. All faded to obscurity in the months and years following.

As Vine fended off smaller competitors, it entered a war with a growing Silicon Valley empire. On June 20, 2013, a few months after he'd downplayed video to the *Verge*, Systrom introduced Instagram Video, which allowed users to create fifteen-second clips on the existing app. Instagram also went beyond Vine's barebones functionality, offering basic editing capabilities and more than a dozen filters specially created for video. "What we did to photos, we just did to video," Systrom said onstage at the Menlo Park launch.

Then a new rival burst onto the scene: Snapchat. The company had launched in the fall of 2011 and in its first version, Snapchat allowed users to share photos with friends that disappeared after being viewed. The notion of ephemerality was embraced by teens and college kids. By October 2012, Snapchat users had shared over a billion photos through Snapchat's iOS app. But Snap's real game-changing feature rolled out in October 2013: Stories. Stories allowed users to share photos and videos to their followers that would expire twenty-four hours after being posted, viewed or not.

Like many of its social media predecessors, Snapchat had not been created for people looking to build audiences. Snapchat co-founder Evan Spiegel was intent on keeping Snapchat's focus on everyday users who wanted to connect with friends, a page out of the old Facebook playbook. But Snapchat users soon began using Stories to micro-vlog their lives, sharing a steady stream of ten-second videos throughout their day. Stories put Snapchat into the same conversation as Vine and Instagram, especially as celebrities increasingly adopted the lower-stress yet more intimate format.

That same year, the video app pack was joined by another one: Musical.ly, the app that eventually became TikTok.

Originally founded as an educational platform called Cicada, Musical.ly was meant to offer students three-to-five-minute videos on various subjects. After burning $250,000 only to learn that the app's target demographic resisted anything that smacked of school, founders Alex Zhu and Luyu Yang pivoted. Zhu had been impressed by the video revolution that Vine had launched. One day, on a train ride through Mountain View, he had a moment of inspiration.

As Zhu observed a group of kids on the train, he noticed that half of the group was listening to music, while the other half was snapping and sharing selfies on social media. What if, Zhu thought, you could combine the two?

Zhu relaunched Cicada as Musical.ly and tapped tech entrepreneur Alex Hofmann to join as CEO of Musical.ly Inc. in the United States. Within a month of their decision to pivot, in July 2014, Musical.ly launched. The app initially functioned as a simple video-editing tool, allowing users to add music in the background of their videos. For about a year, it petered along, essentially ignored as a threat to Vine, Snapchat, and other social media bigwigs. But then, it took off.

※

The fates of tech founders rose and fell throughout this Cambrian explosion of mobile video start-ups, most of which have faded into obscurity. But it wasn't just the Silicon Valley tech circle with a stake in the game anymore.

Each app fostered its own slate of creators. Early Vine users like Jarre and Mancuso shot to stardom at shocking speed. And the same was happening on Snapchat, Musical.ly, and other emerging social apps.

When Snapchat first launched, Shaun McBride was a sales rep for skateboard and snowboarding clothing brands. He was a tall, outgoing guy, with shaggy brown hair and a big smile. He enjoyed his job and regularly zigzagged across the country on business trips during which he'd be away from his six younger sisters. To keep in touch with them, he regularly texted photos of his travels. In 2014, his sisters told him there was a better way. "You should get on Snapchat," they said, explaining that it would be an easier and more fun way to send photos.

When McBride gave it a shot, he took to the format instantly. He snapped photos of his adventures and embraced the app's doodling feature, which let users draw on the photos they took. He started scribbling funny Day-Glo illustrations on top of his images. When his sisters showed their friends, they added him on Snapchat by the dozen.

McBride, or "Shonduras" as he was known on the app, amassed a small following by word of mouth. As people shared screenshots of his snaps on Reddit and Twitter, suddenly McBride became a star. He was among a wave of early Snapchat creators who managed to find internet fame on the app, alongside former teacher Michael Platco, filmmaker Atlas Acopian, and artist CJ OperAmericano.

On Musical.ly, during the summer of 2015, users like Ariel Martin were experiencing the same explosive phenomenon. Martin, fourteen, was stuck at home in Pembroke Pines, Florida, when Musical.ly was experiencing its first huge spike in growth. She had just graduated from eighth grade when a pipe burst in her family's suburban home. She, her two brothers, and their parents were forced to move in with their grandparents for the summer. On top of that, her mother had recently broken her leg, so Martin and her brother Jacob couldn't go anywhere easily.

On long, sweltering days, Martin needed a way to pass the time. Like many teenagers, she turned to her phone. Other kids were talking about a new app, Musical.ly, so she decided to try it out. Rather than use her full name, she chose the handle "Baby Ariel."

Like McBride on Snapchat, Martin developed a style unique from everyone else on Musical.ly. For the first year of the app's life, before Martin joined, teenagers would generally prop up their phone in front of them or hold it straight-on as they recorded their lip-sync videos. But to Martin, this felt stale. In her first video posted to the platform, a lip-sync to "Stupid Hoe," a song by Nicki Minaj, she didn't just hold the camera steady, but instead adopted techniques that had just begun to gain traction on Vine. She swirled her camera up and down and sideways as she mouthed the words, or shook it to match the beat. The result was a chaotic but compelling visual format that would eventually come to define the app.

In retrospect, it seems almost ridiculous that up until that moment, almost nobody had thought to move the camera, but within weeks Martin gained thousands of followers. "It was unbelievable," said Musical.ly's CEO, Hofmann. "She invented the Musical.ly video." The company began featuring Martin's videos on the app, and "all of a sudden, all the other users started to replicate her style," noted Hofmann.

For the rest of the summer, Martin posted a Musical.ly video every single day. "I was getting tons of likes, comments. I was like, 'This is really cool, let me do it again,'" she recalled. With her popularity skyrocketing, she was approached by companies to do brand deals, which she signed. Before freshman year began, Martin was already earning a steady income from deals on the app.

Right before the summer ended, Martin was "crowned," which was Musical.ly's version of a "Verified" badge. "If you got crowned, it was like, 'Oh, you're an official creator,'" she said. "It was huge. I was at my friend's house when it happened, and we were all screaming."

Martin had left eighth grade as an average teenager, but by the time she started high school, she had become internet famous. People whispered about her in the halls and stared; classmates gossiped about her videos. She felt awkward, and making friends felt hard. "Your first year of high school, all you want to do is fit in," she told me. "People looked at me strange."

A month into Martin's freshman year, she was contacted by DigiTour, a popular social media star tour that brought content creators to different cities across the country for meet-and-greets. They were recruiting creators from Musical.ly to tour together and wanted to know if she'd join. Martin had a family meeting about it. She had to decide whether to stay in public school or transfer to an online school so that she could travel the country and focus on content creation full-time.

"Growing up, I never thought of being a creator or influencer as a thing," she told me. "It wasn't like it is now.... No one thought of it as a career where you could really make money and do this as your full-time job. It was all really new and we didn't know." Her summer brand deals were encouraging enough that she decided to take a leap of faith. She said yes to the tour and, that fall, left public school to set out on a journey across thirty cities in America.

"My life went from normal everyday life to the next day being on tour and performing every single day. We'd sit in the back and play video games and talk to each other or look out the window and see the cities pass by," she said. "Each morning we'd have breakfast. It felt like summer camp."

Fans in each city would swarm the group. As the stars collaborated, their audiences grew bigger and bigger. DigiTour boosted their followings, which led to even more opportunities. "At the time, you're a teenager and all of a sudden you're famous on social media," Martin reflected. "It was the coolest thing ever."

※

The spate of new video apps that launched between 2013 and 2015 was dizzying. If short-form video wasn't enough, there was also a sudden boom in live-streaming apps. The notion of livestreaming wasn't new. As far back as the 1990s, dot-com millionaire Josh Harris founded Pseudo.com, which aimed to transform average people into digital VJs. In 2007, entrepreneurs Justin Kan, Emmett Shear, Michael Seibel, and Kyle Vogt created Justin.tv, a website that allowed anyone to broadcast video online, which eventually became Twitch. The interactive video platform VYou allowed users to create video responses to questions posed to them by other community members.

The year 2015 was clearly an inflection point in the live video landscape, opening up a whole new path to online fame for average people. Part of this was due to technological innovation like more processing power in mobile phones, better internet connectivity, and more widespread availability of Wi-Fi, but a lot of it also came down to money. Tech investors and advertisers recognized an exciting new revenue opportunity. "We may have reached a tipping point where live-streaming video isn't just a fun idea, but a viable business with mainstream potential," the *Verge* reported.

YouNow seemed poised to dominate this new market. Launched in September 2011, it initially struggled to grow. But the company introduced a mobile app in 2012 and added features over time that made it easy for anyone to hop on and immediately begin live-streaming.

YouNow's founder, Adi Sideman, had attended NYU's undergraduate film school and then the university's Interactive Telecommunication Program. He was fascinated by interactive media products and recognized the power of online creators early. Sideman believed deeply in what he and others referred to as "user generated media." He saw it as a critical tool for self-expression and connection and had dedicated his career to lowering the bar for average users to create and connect online. "I do think that as the pioneers in enabling live-streaming from anywhere and self-expression in that way, it was answering a need," Sideman told me. "Not just for regular people to connect and alleviate seclusion and boredom and other social matters but [for] creators to connect with their fans in a more meaningful way." YouNow was the ultimate realization of Sideman's vision.

In 2015, YouNow was one of the top-grossing social apps on iOS. More than 35,000 hours of video streamed on YouNow daily, and more than a million dollars in tips flowed through to streamers each month. Most of the people gaining audiences on YouNow were teenagers who flocked to the platform to vent, chat, sing, dance, and connect with each other. Their streams were long. Some creators live-streamed themselves sleeping under the hashtag #sleepsquad, earning money from tips while they slumbered. While Meerkat and Periscope gained traction among adults, celebrities, and the media, YouNow became a monetization engine for a wave of young online creators, especially Musers (as Musical.ly users were known).

※

Gaming had been a big part of the internet going back to the era when alphanumeric characters were the most advanced graphics around. In the 1990s, technological advances allowed players to compete with each other over the internet. In the early 2000s, South Korean gamers began to group together in teams to compete against each other. Tournaments, particularly those involving a game called *StarCraft*, became so popular that professional leagues formed and matches were broadcast on new TV channels dedicated to "e-gaming." Despite the fact that some of the biggest tech companies in America made video games, few believed there was an audience for *watching* other people play games when they could easily play themselves.

Spun off from Justin.tv, Twitch was visited by more than 50 million users per month, a small percentage actually playing games while the rest spectated. (Twitch successfully recruited some YouNow users who enjoyed streaming gameplay.) The star gamers did more than play for an audience; they played *to* that audience, narrating their progress and sometimes rambling about topics that had little to do with the game they were playing. In doing so, they developed personalities that went beyond what Twitch had imagined. An analogue might have been the personality-driven cooking shows that were carving out space on TV and online.

Popular Twitch creators were building real businesses. Imane Anys, known online as Pokimane, leveraged her popularity on Twitch to launch her own massively successful merchandise line. Ibai Llanos leveraged his following to develop a talk show where he interviewed celebrities, from pop stars to professional athletes. Realizing that creators like these attracted users better than any Twitch marketing might, the service tried to make it easier for creators to monetize and provided in-depth analytics so that streamers could measure and improve their reach.

Twitch itself upped the ante when it paired its homegrown stars with other celebrities. A stream of Drake and Ninja, a Twitch creator whose following was in the many millions, playing *Fortnite* together on stream resulted in 600,000 concurrent viewers. Drake's involvement gave Twitch a massive boost in cultural relevance and Fortnite boosted the video-game industry. However, as big as Twitch was growing, it remained fundamentally limited: it wasn't meant for mobile devices, and its emphasis on live-streaming gameplay alienated many.

// **CHAPTER 12** //

Parallel Lines

TWITCH MAY NOT HAVE BEEN ON TRACK FOR A BILLION USERS, BUT IT was a good business. In 2014, after negotiations with Google sputtered out, Amazon bought the platform for nearly $1 billion. Competing apps like Meercat, Periscope, and YouNow battled for live-stream dominion, while Facebook announced a "pivot to video" and sought to buy or copy any video app that showed promise.

In 2014, YouNow launched a partner program that allowed creators to make money from their live-streams. The program transformed the app into a sensation. YouNow had revolutionized live-streaming by making it accessible and mainstreaming the format to Generation Z, many of whom were teenagers and middle schoolers at the time. "[YouNow's] most popular users are largely unknown teenagers," the *Verge* noted at the time. "They aren't broadcasting from exciting places or doing interesting things. A lot of what happens on YouNow feels like the PG-13 version of Cam Girls, part confessional conversation, part vaudeville performance."

Its decision set a new bar for the whole industry. It didn't just push other apps to consider revenue-sharing options; it pushed more creators to diversify their options and become multi-platform talents.

Tayser Abuhamdeh began live-streaming on YouNow in June 2014, using the handle Mr. Cashier. The twenty-four-year-old worked at a bodega in Williamsburg, Brooklyn, ringing customers up for their small purchases like cigarettes, sandwiches, coffee, or snacks. It could get boring, and he figured live-streaming would be a good way to pass the time and entertain himself. He just set up his phone behind the counter and began streaming through his shifts.

"I was talking to myself at first," he told the *Verge* in 2015. "Eventually I started opening up, saying random things, telling jokes and laughing at my own jokes. I started to act like people were there watching, and that's when they showed up."

His early streams were uneventful, just customers coming in and out. Most ignored the stream altogether. Abuhamdeh got more personal, sharing stories from his life. His viewership grew, and he began streaming at home during his off hours.

Eventually, in order to help cover the cost of his phone bill, he joined YouNow's partner program. Before long, Abuhamdeh was one of YouNow's breakout stars. He started getting recognized on the street. All the while, the app took up more and more of Abuhamdeh's time. Soon he was streaming for hours a day.

In 2015, his boss at the deli called to fire him. The boss felt that the live-streams were interfering with his work. Abuhamdeh didn't mind. He took the call from Chicago, where he was snapping selfies with fans. By that time he'd amassed over 135,000 followers and was making triple what he made at the bodega.

※

YouNow had been good to Abuhamdeh, but platform-creator relationships were becoming more complicated. A whole industry of MCNs like Maker Studios and Fullscreen had emerged to foster and manage creators, filling a gap in the platforms' services.

Then, as Vine boomed and YouTubers proliferated, the entertainment industry snapped to attention. After years of dismissing online video creators, they worried they were about to miss out on the next generation of talent. To try to catch up, Hollywood threw Hollywood money at agencies representing YouTubers.

DreamWorks Animation fired the opening salvo when it bought the MCN AwesomenessTV in 2013 for up to a reported $117 million. It was the biggest exit yet for an MCN, and it signified that internet creators could have movie-star value. It also had a domino effect. In March 2014, the Walt Disney Company acquired Maker Studios for a whopping $500 million, with an additional $175 million performance-based bonus. That fall, Otter Media, a web-video joint venture between AT&T and the Chernin Group, purchased a controlling stake in Fullscreen, in a deal valued between $200 million and $300 million.

The money and attention that digital stars were now receiving brought YouTubers, Viners, and other online creators into the mainstream. They went from kids wasting time in their parents' basement to the entertainers of the future.

In the summer of 2014, Shane Dawson and Jenna Marbles, two top YouTubers, graced the cover of *Variety* magazine in a story headlined, "New Breed of Online Stars Rewrite the Rules of Fame." In the accompanying article, Avi Gandhi, then a digital agent at talent agency William Morris, said that he was seeing a growing number of online creators earning seven figures a year, and some were approaching eight. "These digital stars, a lot of them, have online audiences bigger than TV shows," he said.

While the MCN boom earned huge payouts for their founders, MCNs' lasting value was questionable. Once a YouTube star became big enough, they didn't really need an MCN. They could instead work directly with their own personal manager, like Keyboard Cat's Ben Lashes, and set their own rates. Thus, mid-sized and smaller YouTubers became MCNs' bread and butter. But they, too, realized they were forking over a significant portion of their ad revenue to MCNs and for dubious benefit. As the YouTube economy thrived, something rotted within MCNs.

"The investors were looking to capitalize on their investment. There was this emphasis on growing fast, because if we didn't we'd be left behind," said Kassem Gharaibeh of the Station and Maker Studios. "It stopped being about making this cool studio where creators could come and execute their creator vision, and it became about money."

Each MCN signed as many creators as possible—tens of thousands—in the hopes that a few would become superstars. Maker alone had 55,000 content creators in its network by 2014, and unsurprisingly it could no longer provide high-quality, personalized services to every one of their creators. "We had so many channels that unless you were among the bigger ones. . . . you didn't see a whole lot of benefit to signing over your channel and giving up a portion of your revenue," said Gharaibeh.

Two years after the Disney acquisition, Maker cut its creator network from thousands of channels to three hundred and eventually laid off much of its staff. The company was absorbed into Disney Digital Network in 2017. Machinima also made significant cuts in the mid-2010s and shut down entirely by 2019. Fullscreen eventually sold its remaining stake to AT&T and the Chernin Group and Strompolos left the company to focus on new ventures.

※

As the mainstream online creator universe evolved, a dark parallel universe had emerged beside it. The internet was becoming more radicalized, but extremist ideology had, for years, been relegated primarily to more obscure corners like message boards and subreddits. In August 2014, all that changed with a series of events known as Gamergate.

Gamergate was a coordinated, deeply misogynistic harassment campaign that relied on manufactured outrage cycles to terrorize women who espoused progressive values. It began when the ex-boyfriend of video-game developer Zoë Quinn attacked her in a blog post, making false and defamatory claims about her that were then elaborated upon and amplified by anonymous Redditors and 4chan users. Several other women became characters in what was an increasingly complex fabrication driven by antagonism toward advocacy for greater diversity in the video game industry, and then just general antagonism toward liberal ideology. Thousands of men used the hashtag #gamergate to identify and target outspoken women in gaming with nonstop rape and death threats.

Nastiness was not new on the internet, nor was sexism (or racism, or antisemitism, or homophobia, or just about any malicious form of discrimination). Gamergate was a watershed moment in online culture because it provided a blueprint for the weaponization of social media platforms to spread hate, exploiting the media's ignorance of online culture and tendency to take things on the internet at face value. Conspiracy theorists crafted a bizarre narrative in which liberal forces were in league with the media to destroy America. The perpetrators relentlessly mocked their targets with odious memes, tweets, and posts. They sought to get women fired by making bad-faith attacks to the women's employers in a strategy that eventually became right-wing standard operating procedure. These tactics have been used in the near-decade since to push countless women out of their jobs in journalism and the gaming and entertainment industries.

Gamergate also catapulted a new type of influencer to fame. The alliances that formed among Gamergate's participants, including Milo Yiannopoulos, Candace Owens, and other future right-wing internet stars, would metastasize to terrible effect. In their book *Meme Wars*, Joan Donovan, Emily Dreyfuss and Brian Friedberg argue that it was Gamergate that created the opportunity to funnel "young reactionaries to MAGA." Steve Bannon, central to Donald Trump's election, agreed, praising Yiannopoulos in particular for creating, through Gamergate, the "army" that got "turned on to politics and Trump."

These far-right online creators, and the groups of reactionary influencers who followed in their footsteps, eventually came to amass great power online, serving as the backbone for the broader right wing media ecosystem.

❉

Vine stars, however, were operating in a separate online universe. As Jérôme Jarre saw his follower count soar in early summer 2013, he realized that few of his peers knew much about business. He was confident that he could help Vine stars connect with brands, but he needed a partner. He cold-called Gary Vaynerchuk, the entrepreneur-cum-influencer who had risen to prominence by leveraging YouTube and Twitter. Jarre didn't have Vaynerchuk's number, so he came up with an audacious plan. As Vaynerchuk tells it, he was holding an event in Toronto. During the Q&A, Jarre took the mic to challenge Vaynerchuk to a game of rock-paper-scissors. If Jarre won, he proposed, Gary Vee would get coffee with him.

Jarre won the game. At coffee, he persuaded Gary Vee. Jarre laid out his vision for a new company, and the pair agreed to start a new advertising agency that would connect Vine stars with brands. "It took him all of seven minutes to convince me to start an agency with him," Vaynerchuk told the *New York Times*. "He understood Vine, Snapchat and Instagram in a way that no one else did."

In May 2013, the duo founded GrapeStory. The model of matching sponsored deals with blue-chip brands to creators was by now familiar in digital media, but no one else was working with Viners yet. Jarre had access to all of them.

One of GrapeStory's first deals was a General Electric campaign featuring Jarre and Marcus Johns in zero-gravity. The clip shot to number one on Vine's Popular page. The agency partnered with top Viners like Logan Paul and Nash Grier. By early 2014, Grier was able to command $25,000 for a six-second ad, the equivalent of $4,166 per second.

Soon came other rival firms. Niche co-founders Darren Lachtman and Rob Fishman had followed the rise of YouTube's MCNs closely. Lachtman had recently worked at Bedrocket Media Ventures, a company that oversaw its own network of YouTube channels. Fishman was an early employee at the *Huffington Post* and had watched Arianna Huffington build her personal blog into an empire. The pair initially batted around ideas related to YouTube, but given the saturation of MCNs, building another one didn't seem appealing. Vine, on the other hand, was wide-open. The company wasn't linked into its parent Twitter's ad network. So, if brands wanted to run campaigns on the platform, they had to go directly to the talent.

Vine stars commonly listed their contact info in their profiles, and conveniently many were based in New York. Lachtman and Fishman kept their day jobs and at night began meeting up with Mancuso and Johns, asking, "If we brought you brand deals, would you do them?" Lachtman recounted. The pair also cold-emailed the top Viners asking the same thing. The answer from the Viners was a resounding yes.

Soon, Niche had landed its first client, CBS Films, which ran a promotion where Viners Brittany Furlan, Johns, and Mancuso posted advertisements for CBS programming—in the form of Vines—on their own accounts.

From there, business took off. Vine itself grew rapidly over the course of 2013 and 2014. Digital media sites like Mashable, BuzzFeed, and the *Huffington Post* aggregated viral content from the platform, repackaging Vines into viral roundups and meme explainers. Vine even reached as far as the White House, which used it frequently to document life at 1600 Pennsylvania Avenue.

The Niche team caught Twitter's eye. They formed a partnership where Twitter reps would refer brands interested in sponsoring Vines to Niche. "Any time an advertiser at Twitter did a big deal with Coke," Lachtman said, "Niche became the de facto pass-along."

These big campaigns led by Niche heightened the legitimacy of Vine in the eyes of marketers. One campaign, for computer giant HP, featured Viners Jessi Smiles, Brodie Smith, Robby Ayala, and Zach King sharing an HP notebook/tablet. The ad was essentially a supercut of Vines produced by the creators, and it ran as a TV commercial during the World Series. The gum brand Trident collaborated with Mancuso and Megalis on another TV ad, where they crooned about the gum's "layers of flavor" while holding a giant pink teddy bear. Logan Paul traveled across the United States and completed dares posed by his followers as part of a promotion for the underwear line Hanes.

Niche was facilitating so many deals by the end of 2014 that Twitter stepped in and offered to buy the company outright for around $50 million in cash and stock. The Niche acquisition felt like a breakthrough to Viners. Niche was doing $10 million in revenue the year Twitter purchased it, and the following year it grew to $35 million. But the deal meant more than money; it indicated to creators that companies beyond YouTube treated their work as legitimate and valuable. Maybe, just maybe, Vine itself would see the light.

On average, by 2014, advertisers were paying talent around $3 to $5 for every thousand followers, according to Vine creators. A Vine star with 1 million followers could earn $5,000 for posting a sponsored Vine to their feed. *Recode* reported that former waiter-turned-Vine-star Matt Cutshall, who had (only) 323,000 followers, was making close to $75,000 a year doing advertisements for brands on social media including the Gap.

The biggest creators on Snapchat were finding similar success. Consumer goods and food brands like Taco Bell, Chipotle, and Disney embraced this approach, often working with new agencies like Carrot Creative to set up accounts and publish daily snaps to the brands' Story feeds.

As with early Instagram, however, only a small slice of top Viners and Snappers had enough followers to make a healthy living through sponsored content alone. Even accounts with hundreds of thousands of followers needed other forms of income. Where that income would come from was an open question.

Some rising creators, especially young ones, thought they might have an answer. It had everything to do with what Mancuso and Jarre had experienced in Central Park: Their fans treated them like stars—so why not start to *act* like stars?

※

In early 2013, a Louisiana-based entrepreneur named Bart Bordelon had a eureka moment. While shopping at a local mall, he witnessed a meet-up for Aaron Carpenter, a fifteen-year-old Instagram star. Hundreds of teen girls flocked to the mall for a chance to meet him.

For Bordelon, the experience was eye-opening. Kids like Carpenter had passionate fan bases, but they were left to organize their own no-budget events at which they were mobbed. Bordelon realized the events industry was sleeping on social media.

Bordelon got in touch with Carpenter and asked what he thought about replicating the event, but with major upgrades. Bordelon pictured a bigger stage, more professional production, more showmanship, and more of a crowd. Carpenter was intrigued. The two began brainstorming and thought of inviting other up-and-coming creators. They also came up with a name: Magcon, short for "Meet-and-Greet Convention." Bardelon's vision was to gather a critical mass of social media talent and send it on tour, turning out legions of screaming fans in cities across the country.

Magcon wasn't the first live event aiming to tour creators. A few years earlier, in 2011, music manager Meridith Valiando Rojas and her then boyfriend, Christopher Rojas, launched an event called DigiTour. Officially backed by YouTube, the first DigiTour featured a range of YouTube music creators like The Gregory Brothers, DeStorm Power, and Joe Penna. (Penna's channel, MysteryGuitarMan, had become one of the most-subscribed channels on YouTube in 2011.) DigiTour grew to include YouNow stars and Musers like Baby Ariel, as well as top creators from Vine, Twitter, Instagram, and even some of the smaller platforms. Along with traditional musical acts, the 2013 tour would feature up-and-coming talent like Zoella, a U.K.-based beauty creator.

The DigiTour creators performed skits and sang across major cities in the United States The first tour ran for six weeks in twenty-seven cities in North America, with thousands of teens and children assembling in snake-like lines to get posters signed and meet their heroes. General admission to DigiTour was around $25, but dozens of add-ons were on offer. And $75 extra got you a selfie with a creator.

The timing for a DigiTour competitor couldn't have been better. DigiTour was focused on YouTube, but Vine was growing exponentially, minting scores of new influencers who had more fame than they knew what to do with. With Magcon, Carpenter and Bordelon sought to tap into that energy, working together to scout prospects, largely in the south where they were based. With a target audience of teenage girls, their primary criteria for talent were youth, handsomeness, and beaucoup de followers.

Their biggest signings were two floppy-haired teenage heartthrobs: Nash Grier, a sixteen-year-old from Davidson, North Carolina, and Cameron Dallas, a nineteen-year-old from San Bernardino, California. Both had blown up on Vine in the app's early days. The tour would only drive their follower counts higher.

The first Magcon tour kicked off in Houston in September 2013. It operated similarly to DigiTour, featuring short performances and brief meet-and-greets and photo-ops. Fans paid rates from $30 to $150 for a range of perks. The full roster consisted of Aaron Carpenter along with Vine and YouTube creators, including Cameron Dallas, Nash Grier, Shawn Mendes, Matt Espinosa, Carter Reynolds, Taylor Caniff, Jack and Jack, Nash's brother Hayes, and one woman, Mahogany Lox.

Magcon was a hit. Onstage, the creators performed karaoke, goofed around, and traded barbs. They'd been selected for their follower counts and looks over singing or performing abilities, which made their goofy, amateurish performances amusing and endearing. The crowd's passion was nonetheless intense. Every stop on the tour was bigger than the last.

In past eras, teen stars arrived through the music industry, movies, or TV. They were trained and vetted. They had put in thousands of hours and jumped through scores of industry hoops. Social media, by contrast, didn't impose such a gauntlet, and fans didn't seem to care. The performers were chased through the streets of the cities they visited. They could hardly do media appearances. When word spread through fan-driven Twitter accounts that Dallas was coming to the *People* magazine offices for an interview, young girls swarmed the building and security was called.

Magcon encounters fed the viral beast. Snapping a selfie with your idol was a priceless memory for the fan, but also, when posted online, valuable promotion for the star. Fans repackaged moments from the tour in Vine mashups and YouTube compilations that went viral. A video titled "Funny moments Magcon boys Pt 1," for instance, garnered over 2.7 million views. A compilation of Magcon-boy Vines came close to 5 million views.

Magcon found early success, but it didn't last long. The tour had received pushback from parents and unsatisfied fans who felt that the event didn't give them their money's worth. "If you shell out a bunch of money for a One Direction concert, you know you're going to experience a massive show," *Business Insider* reported. "But the internet stars of Magcon and their managers were charging young fans and their parents big bucks to stand in line for a selfie and see a dance performance comparable to a high school talent show." Just eight months after it began, Cameron Dallas, Nash Grier, Hayes Grier, and Carter Reynolds, some of Magcon's biggest stars, announced they were quitting. They didn't feel the tour was good for their brand anymore; they wanted to be seen as individuals and focus on their own careers.

Even so, no one was prepared for the reaction when Magcon announced that it was ending. Less than an hour after the news broke, seven of the top ten trending hashtags in the United States were related to the end of Magcon. #RIPMagcon and #thanksbart went viral, along with tributes on Vine and YouTube. Young girls posted tribute Vine compilations and videos of themselves weeping over the end of the tour. Trolls seized the moment and mocked young people who were openly discussing suicide following the Magcon breakup. The hashtag #cutformagcon trended nationally, viciously encouraging the tour's young followers to self-harm in an effort to force Cameron Dallas and other social media stars to rejoin the tour.

The outpouring of support inspired the tour to return the following year, albeit without several of its primary headliners. Dallas rejoined for a follow-up European tour—also marked by myriad screaming fans and event-planning woes—which was documented in the 2016 Netflix series *Chasing Cameron*. But Magcon would never again reach the heights it saw in 2014.

Building on the Digitour experiment, Magcon had reset the equation. As in the music business, touring became an integral part of the creator business model, especially for young stars who'd emerged on Vine and other mobile apps. Soon there were events from MCNs like Fullscreen, which launched a Girls' Night In Tour. The restaurant chain Buca di Beppo hosted a Magcon-like tour where young viral stars hosted meet-and-greets and shows at a series of the chain's locations.

DigiTour itself expanded aggressively, selling 120,000 tickets in 2014 and raising $2 million from investors including Ryan Seacrest and Condé Nast parent company Advance Publications. Just a year later, DigiTour doubled ticket sales to over 250,000 and raised another $10 million from Viacom and other investors. The company also put together custom tours for creators like Jack and Jack, a YouTuber duo who had been part of the original Magcon, as well as O2L, a massively popular collab group. In 2016, DigiTour evolved to recruit top stars from all major apps—not just YouTube—including Instagram sensation Blake Gray, Musical.ly's Baby Ariel, and YouNow stars Weston Koury and Nathan Triska.

But by 2018, it, too, was struggling. "There was so much fragmentation with tours," said former DigiTour CEO Meridith Valiando Rojas. "It got to the point where it became too many, and it was diluting the market and it lost that special sauce."

While the tour boom ultimately proved to be an artifact of its time, the crowds that creators generated were undeniable proof that a major audience existed for online influence. As more online creators were embraced by the entertainment industry and seen as celebrities in their own right, the rates they could charge for things like a cross-country tour became so high that touring them as a group would have been nearly impossible. For creators, this was, in many ways, a victory. But for fans, it felt like a defeat. The possibility that you could manifest the internet on a single stage, and then meet, embrace, and hang out with your idols, had evaporated.

There was also a darker, sobering reason why tours grew more expensive and untenable: the increasingly undeniable need for serious security measures. In 2013, creators were happy to wander into malls with no security and get mobbed by fans. But that was a naïve dream years later. The turning point came in June 2016, when the YouTuber and music artist Christina Grimmie was murdered by a fan in Orlando. Grimmie had posted on social media asking people to attend a concert she was giving and, after her performance, she stuck around to meet her followers. As she opened her arms to give a twenty-seven-year-old fan a hug, he fatally shot her once in the head and twice in the chest. Police later revealed that the fan had stalked her for years online and fantasized about the two of them being together. When he realized he couldn't have her, he chose to murder her. Top creators at future events like VidCon, DigiTour, and Playlist Live were given security escorts, while access to creator lounges became more restricted. VidCon also installed metal detectors and banned all informal meet-and-greets.

※

The tour era is still remembered fondly by Vine stars like Cameron Dallas and Nash Grier and their fans. But the investors who bet that touring would prove an essential model for online creators lost out.

The creator class was doing what it had done each time it had been faced with constraints: develop new revenue pathways. This time creators sought a revenue stream that didn't require exhausting travel. Their solution would transform the industry yet again.

// CHAPTER 13 //

Counting Seconds

By 2015, Vine stars had come a long way. Like the first class of YouTubers before them, Viners were now working with agents and managers. To transition from internet novelty to lasting star, they went to Hollywood. Literally, they moved to Los Angeles. They took a page out of the Station's book, and top Viners moved into an apartment building on the L.A. street that had their name on it: 1600 Vine.

Collab houses were not new in the online creator world. In the aughts, bloggers Ezra Klein, Brian Beutler, and Ben Adler lived together in the "Hobart House" in Washington, D.C., so called because it was on Hobart Street. There was also a D.C.-based "blog house" called the "Flophouse," where the writers Matthew Yglesias, Kriston Capps, and Spencer Ackerman lived and blogged together. In the Viners' case, the massive luxury apartment complex at 1600 Vine served as the epicenter of the Vine era of the internet. Everything happened there: massive video shoots, star-studded gatherings, life-changing business deals. Eventually, 1600 Vine would also be the scene of the app's demise. But, for a time, the place was a prelapsarian paradise.

Brent Rivera, a Vine star who went on to YouTube and TikTok stardom, said the perfect collab house should "be big, and the more amenities the better, like a pool, nice bathroom, nice lighting, big back-

and front yards, room for activities and fun stuff you can do inside or outside." It should also be removed from prying neighbors, but still somewhat centrally located.

Andrew Bachelor, known online as "King Bach," a charismatic twenty-five-year-old, was the first to move into the complex in early 2013, before he downloaded Vine. "The only reason I picked 1600 Vine is because I used to live on a street called Vineland in Studio City," he said. "I moved to 1600 Vine because I liked the Vine name." A year later, he had become one of Vine's biggest stars through his signature style of humorous skits, parodying "relatable" situations like missing a friend's high-five, going to extreme lengths to keep his sneakers clean, and dealing with racial profiling as a young Black man.

Once Bachelor had made a name for himself, he met John Shahidi, a well-known Hollywood businessperson who at the time worked closely with Justin Bieber. Shahidi was running the app Shots, a selfie-focused photo app co-founded by Bieber in 2013, and he wanted to get more Vine stars to use it. Shahidi had been toying with the idea of getting a place in L.A. for a while, and after visiting Bachelor at 1600 Vine, he signed a lease for one of the penthouses. His brother, Sam Shahidi, also a well-known entrepreneur who co-founded Shots with his brother John and Bieber, moved in too. "Shots was trying to be the photo service for creators, whereas Vine was the video service for them," Shahidi said. While the Shots app wouldn't survive, Shahidi's investment ended up luring scores of Vine stars to the building.

Shahidi's 1600 Vine two-story penthouse had tall windows and lots of light. He lived in one of the bedrooms downstairs, while the upstairs served as the Shots HQ. Bieber would come by frequently for meetings and just to hang out.

Bachelor, eager to work a celebrity into his content, started making Vines with Bieber in the building. Other Viners took notice, and they began to come over to collaborate and make videos with Bieber and hang out at Shahidi's penthouse. The building had everything a content creator needed to create the perfect video: endless hallways, common areas, a pool, a workout room, and more. The large, modern apartments with their big white walls offered perfect blank canvases for videos. 1600 Vine was also centrally located. It was easy to step outside and prank unwitting bystanders or film videos at nearby landmarks.

At this point, Randy Mancuso had been bouncing around L.A. for a year, creating content while also pursuing his dream of becoming a film director. At the end of 2014, he looked into the building where Bachelor was making his content, found plenty of open units, and moved into one. With Mancuso, Bachelor, and Shahidi there, other Viners followed. Logan and Jake Paul got an apartment together at the complex, and then came Lele Pons, the most-followed woman on the app, along with a slew of others. Anwar Jibaw, another top Viner, moved in and became a favored collaborator of Bieber's. By mid-2015, nearly twenty of the app's top creators lived in the complex. "It felt like a bunch of dorms although we were young, stupid, and rich," said Mancuso. "It fulfilled the college experience I didn't have. It was the college experience on steroids."

Lele Pons had just finished high school when she moved into 1600 Vine in 2015. She also viewed it as a college experience. "I was like, 'Let me go to Vine college,' basically," she said. "Everyone was in that building and we all lived and collabed every single day." Pons's mother became something like the house mom of 1600 Vine, living with her daughter in a spacious two-bedroom apartment.

While life was chaotic in the building, there was a rhythm. Each day, creators woke up around ten a.m., sometimes hungover. They'd check Rankzoo, a platform that offered regularly updated rankings of Viners based on a variety of metrics. Rankzoo was owned by Collab, another early multichannel network, which was founded in 2012 by three brothers, Tyler, James, and Will McFadden, and Soung Kang, an early leader at Fullscreen. After breakfast, creators would head down to the pool to strategize about what to post that day. They'd toss out skit ideas and punch lines. Many skits required several people, and they all took turns acting in one another's videos or holding the phone to shoot. It was like an internet-driven theatre company. "It was so easy to film, I could knock on Amanda [Cerny] or Logan's door and shoot a video," Bachelor said.

Once they had the day's content planned, the 1600 Viners discussed distribution. The top Vine stars practically controlled what was popular on the app. By using the "revine" feature, which allowed you to repost someone else's Vine to your followers, the most-followed users could dominate the app's Popular page. (Top Viners even found an additional revenue stream by charging up-and-coming creators for revines.) This dynamic didn't emerge on platforms like YouTube and Instagram at the time because there was no native reshare tool. Vine was uniquely vulnerable to reach monopolization by a collective of top creators.

On an average morning at 1600 Vine, the Viners would work out who would share each video, as well as when each piece of content would be shared for maximum impact. Bachelor was growing the fastest, with nearly 100,000 followers a day, so he often took the lead. Making an appearance in one of his videos could be a major break, resulting in thousands of new followers.

The creators at 1600 Vine weren't just gaming the app's ranking algorithm; they were pushing the boundaries of the app itself. Vine used to remove any content that appeared to be created outside the app, but when Marcus Johns posted a video celebrating his 1-million-follower milestone that was clearly edited on a third-party app, the company didn't remove it. Other creators saw this and began posting pre-edited content to their feeds as well, adding sound effects or funny visual details to make their content pop.

Jeremy Cabalona was Vine's first non-design or non-engineering hire, joining the company from Mashable in September 2013, where he had been its social media editor. The concept of a "social media editor" or "social media producer" who oversaw a brand's presence on social platforms had just begun to take hold. BuzzFeed, *Huffington Post*, and other upstart digital media companies were among the first to create full-time social media editor positions, jobs that were largely dismissed by leaders in most legacy newsrooms. Social media editors were generally people in their early twenties, often a step above an intern. Yet these entry-level employees often had full liberty to post autonomously on behalf of their companies. Cabalona, himself a recent college graduate, had started Mashable's Vine account the day the app launched. He had successfully programmed it with silly humor and short clips from Mashable's office in Manhattan's Flatiron District. The media company even created a dedicated "Vine studio" in its New York newsroom.

At Vine, Cabalona was tasked with running the company's social media channels and off-platform marketing, as well as growing and engaging with Vine's community. "The company's relationship with creators was nonexistent," he quickly realized. "I would tweet at them, tag them, and share their work, but the overall policy from the founders was to ignore them. There was a bit of resentment that these creators had taken over their network. They were like, 'We don't like this, this is not what Vine is for. . . . It wasn't meant to be a performance platform or to get famous, it was to capture your life moments.'"

Flare-ups between top Viners and the company continued throughout 2013 and 2014, even over the smallest features. In December 2013, Vine introduced web profiles and vanity URLs, short, easy-to-remember links that allowed users to share their Vine profiles outside the app. Top Viners said they received no advance notice of the feature's rollout, however. Nash Grier, one of the Magcon boys and one of the biggest stars on Vine, was minutes too late when the URLs launched. A random user had seized his URL, vine.co/nashgrier. Frustrated, Grier reached out to the company to ask for help in obtaining his username. It would have been simple enough to fix, said Cabalona, and he asked the founders to help Grier out. But Vine co-founder Colin Kroll scoffed at the request. "Colin was like, 'He doesn't need it, let it go,'" Cabalona recalled. "It was a pretty explicit no from Colin." Grier claimed the account "griernash," but it caused unnecessary confusion and harmed his search discoverability on the platform for years.

Tensions between top Viners and the company grew. Creators like Grier realized that the company was silently barring some of their clips from trending. Creators were spending hours and hours on the platform, and they came to know its patterns well. Rule of thumb: 1,000 likes won a spot on the "Popular" page. When content well above this threshold suddenly disappeared from the Popular page, they realized something was up. Often, Cabalona was their first point of contact. "The creators were so in touch with how the feed worked and how the app worked," he told me. "They'd be like, 'This is what numbers I'm doing, so I should be at this spot, but I'm not.' They'd reach out and say, 'Why isn't this video on there?' People would reach out to me all day and be like, 'You're blocking us, we know it.'"

They were right. The co-founders were curating what trended on the site. If a Vine star was rising on the app's Popular page with a juvenile prank or a video that the co-founders didn't approve of, the executives would have it pulled. "Someone would say, 'This kid is trending. Knock him off,'" said Cabalona. Making matters worse, Cabalona wasn't allowed to acknowledge the practice externally, which eroded trust between creators and the app.

By August 2014, Vine began allowing all users to upload prerecorded videos. The feature drew Vine away from its original concept. As its predecessors had done, Vine broke its own rules only after users had bent them. Competitors like Socialcam, Instagram, and Musical.ly had realized this too. But Vine's leadership resented it.

※

In early 2014, tensions over the app's direction came to a head, coinciding with several executive departures at Vine. Jason Toff, a product manager at YouTube, joined Vine as head of product. He brought a different attitude toward Vine's stars. Whereas Kroll had felt there should be an "invisible wall" between the company and its creators, Toff wanted a warmer relationship with power users. "I felt like we had to engage these people who were core to our platform," he told me. "My strategy was to embrace that rather than fight it."

Toff spearheaded several innovations to make life easier for both creators and the company's internal teams. The app introduced a "loop count," publicly displaying how many times a video had been watched or "looped." This provided creators and brands with a new metric by which to judge content success and account reach. Toff also rolled out sweeping changes to the "explore" page and dedicated engineering resources toward tools for Vine's editorial team, allowing them to better curate trending content.

When VidCon 2014 rolled around, it should have been Vine's coming out party. Now in its fifth year, VidCon attracted thousands of teenagers to the Anaheim Convention Center in California for its annual gathering of internet stars. Vine sent a handful of team members, including Toff and Cabalona. While the top Viners were mobbed by screaming fans, Vine's modest booth handed out rubber bracelets that read DO IT FOR THE VINE. Meanwhile, the reigning king of online video YouTube had blanketed the event with branding and sent dozens of staffers.

If the Vine team expected warm greetings from creators, they were brutally disappointed. As word spread that Vine employees were at the convention, Viners grew indignant. When creators encountered Vine staffers, they heckled them. At one point, Cabalona and a colleague fled a particularly contentious exchange with a group of Vine stars out of fear of physical assault.

Afterward, Vine tried to improve creator relations: They hired André Sala, a former digital director at Showtime, to oversee content and partnerships. Cabalona did his best to represent Viners' concerns to management. But the company's rocky relationship with Twitter stifled efforts to hire new people or launch its own partner program.

At the next VidCon, instead of ignoring its star power, Vine leaned into it. The company spent $20,000 projecting footage shot by Cabalona and top creators onto a massive screen. "We shot promo footage with them and tried to make them the faces of the app," Cabalona said. The promotion was the first time Vine had ever featured talent from the app in its marketing. The company launched the campaign through a new Vine account called "Viners."

Vine also sought to impress its own stars by throwing them a celebratory party at VidCon. Inviting top talent, the leadership team hoped it was the prime opportunity to connect with them. Vine employees stood at the app's booth on the main floor of the convention and networked with creators who stopped by. They hosted unofficial meet-and-greets where talent could connect with fans.

But for all the money Vine put into the event, the company still didn't seem to understand how to make the most of those relationships. An entire year had passed since Cabalona had scurried away from furious Viners, and a year was a lifetime when it came to social media.

✳

In the fall of 2014, YouTube spent millions flooding the market with a nationwide ad campaign featuring top talent like Michelle Phan, Bethany Mota, and Rosanna Pansino. YouTube plastered billboards nationwide with the faces of its stars, featuring them in TV commercials and print ads. "When it comes to promoting YouTube," the *Wall Street Journal* reported about the campaign, "Google's strategy is simple: tell the world just how popular some of its program creators really are." Vine wanted to do the same, but it was too late. Many Viners were now broadening their posting habits, releasing content on YouTube, Instagram, and newer apps to diversify their options. They were drifting away from Vine.

In the aftermath of VidCon, the company brought in a new hire to formalize relationships with creators. Karyn Spencer had previously run social media for high-profile celebrities like Ashton Kutcher, building him into one of the most followed accounts on Twitter. She had also worked on the monetization side at the Audience, an influencer marketing firm, where she negotiated brand deals with many Viners. At Vine, her job was to be a peacemaker, building a bridge between the most popular users on the app and the company.

Spencer started in September 2015 with a lot of optimism, but soon grasped the magnitude of the task before her. "Three weeks into my employment they called me in and shut the door and were like, 'The reason you're here is because we're in code red,'" Spencer told me. "That's when I found out my shiny new job actually had a precarious remit attached to it."

Vine's numbers began to crater. Less than a month after Spencer joined, Twitter laid off Vine's last remaining founder, Rus Yusupov, who had served as Vine's creative director. With Vine's founders now all gone, some at the app hoped they could develop a deeper relationship with Twitter. Unfortunately, Twitter was not in a good spot either. Yusupov was laid off among nearly three hundred others across the company.

Spencer, jarred by Yusupov's abrupt layoff, decided to view the changes as an opportunity to repair the company's long-standing rift with talent. She started by meeting up with Viners at events and arranging opportunities for them. In October 2015, she brought a group of them to meet with First Lady Michelle Obama in hopes of inspiring the creators to produce content related to Obama's campaign to encourage higher education.

Spencer also devised other ways to engage the app's top stars. "One thing I observed working in traditional Hollywood was the whole notion of fancy Hollywood events," Spencer said. "Award shows ride on the attendees. If a Hollywood's women's association wants to have a well-attended lunch, for instance, they'd say to Demi Moore or someone, 'Hey, we're going to present you an award for the most powerful women,' for example. As a celebrity you feel honored, and that association uses your name."

Spencer could see from Rankzoo and internal analytics that Andrew Bachelor was consistently the top Viner. She reached out to his sister, who was his manager, and told her that the company wanted to name him "Viner of the Year."

"I'm the new head of creators," Spencer told Bachelor, "And I want to congratulate you and throw a party in your honor in L.A. and work with you on the guest lists, so all of your friends and contacts attend the party."

Viners were skeptical. "Every time I reached out to a Viner, I had to go through this series of verifying who I was, because no one at Vine had ever reached out to these people," Spencer said. "Bach said to me, when we first sat down, 'Oh, I always assumed Vine hated Black people because they never responded when I attempted to contact them.'"

The Viner of the Year party was held on November 9, 2015, at a large, modern rental house in Venice Beach. Nearly every top Vine star showed up to congratulate Bachelor and see what the deal was with Spencer. It was the first big party Vine held for creators, and the house was stocked with Vine-themed merch, balloons, and napkins. There was a photobooth for shooting Vines, props at the ready, and a cash prize of $1,000 for whoever created the best Vine at the event. A large screen displayed Bachelor's popular Vines on loop, and there was even a trivia game people could play with facts about Bachelor's life.

The party went late into the night, with dozens of Viners crowded on the rooftop drinking, eating, and creating content. It was the first time some creators had met in person, and there were lots of hugs and happy tears throughout the night. Spencer was thrilled with the outcome. Viners seemed appreciative. Things seemed to be turning around.

But toward the end of the evening, Marcus Johns pulled Spencer aside and asked if she and others at Vine would be in town the next day. They wanted to invite Spencer and her team to 1600 Vine for a meeting. She said of course, but was nervous about what they planned to discuss.

※

Spencer knew that cultivating top Viners was crucial to the company's survival, but she and many others at the company were conflicted about the creators at 1600 Vine. "They all formed an alliance and any time one of them would post a Vine, they'd drop a link in the group chat and every other creator would revine their vine so it would be the most successful on the app," Spencer noted. "They built this structure of power that kept them all at the top." This collusion made it difficult for users to discover new talent, and for Vine to nurture that talent.

This monopoly, in Vine's eyes, was especially problematic because the content that the 1600 Viners produced was not what the company wanted to promote. Their jokes were often juvenile at best, and frequently misogynistic, homophobic, or racist. "It was a lot of slapping women on the ass and racially fueled jokes like 'white people be like' and 'Black people be like,'" said Spencer. "It was the lowest common denominator of comedy." In one Vine, for instance, Bachelor tells a group of men and women that for him to hold the door to an elevator, they'd have to perform a sexual act. The group laughs, enters the elevator, the door closes when Bach says, "I was serious, though." He then begins to hit the women and men inside. In the final scene, he emerges from the elevator, zipping up his fly while the young men and women lie in various states of undress. The punch line was that he sexually assaulted them. And this was the most popular Viner.

Videos like Bachelor's seemed to support Vine's long-standing efforts to remove content that it found offensive. But because the platform was dominated by a single cohabiting clique, the company had nowhere to turn. If it pulled Bachelor's video, would he defect to another platform? Would all of 1600 Vine defect in solidarity? It was a dilemma that would plague social media platforms for years to come: What do you do when your most popular creators behave badly?

As she threw the Viner of the Year party, Spencer wondered what the app would look like without 1600 Vine's stranglehold on the "Popular" feed. Was this type of content simply being force-fed to users by the 1600 Vine mafia or did users inherently crave it? Would it be possible to off-ramp the most objectionable 1600 Vine creators—instead of compromising to keep them—and give a fighting chance to emergent creators who posted more diverse content?

These questions were lurking in the back of Spencer's mind when she ventured to 1600 Vine the morning after the big party. With her were Jeremy Cabalona and William Gruger, a content strategist who had joined the company from *Billboard* to help Vine develop closer relationships with the music industry.

When they arrived, Johns, the leader of the meeting, ushered them into a conference room where about twenty of Vine's biggest stars sat around a table. The Vine team was eager to hear what they could do to make the creators happy. They were well aware, as Gruger said, that the Viners "had this oligopoly over viewership on the platform." With the app's usage fading, they knew they needed most of the 1600 Viners on board. But rather than being a traditional business meeting, it morphed into a therapy session. Creators vented about feeling neglected. Each one launched into soliloquies about everything they'd achieved thanks to the app, such as buying their family members a car or a house or selling a pilot to a major TV network. "*But*," said one Viner bitterly, in a sentiment that other creators echoed, "you people are the first people I have ever met that work at Vine."

Johns added that while driving down the streets of Los Angeles he saw massive billboards featuring YouTube stars. "Why can't we be that for you guys?" he asked angrily. Where was their recognition? Twitter simply did not care about creators the way YouTube did, he alleged. And no one in the room, including the Vine employees, knew how to change the parent company's core attitudes.

Creators weren't just angry about monetization; they were frustrated that the app seemed to have no interest in building features that catered to power users like them. One issue was harassment. Jon Paul Piques and other Viners felt that Vine and its parent company were not doing enough to curb the brutal and overwhelming abuse directed at top creators, especially women. Creators demanded that Vine institute comment filtering, blocking, and muting features, and more robust analytics.

Spencer attempted to quell the crowd, telling the creators that she heard them, that she understood what they were going through. She told them that although she was new, the very existence of her role and her team meant that the company was turning over a new leaf and was now ready to embrace the importance of Vine stars. She said Vine wanted to find a way to move forward and get a fresh start.

Johns responded coldly. He said that they weren't interested in lip service. They knew how much it would hurt the app if they all stopped posting. He explained that Vine was at their mercy and that in order for the creators to continue to post, the company would need to adhere to a list of demands they'd prepared.

The primary demand was $1 million a year for three Vines every week.

Johns said they'd spell out the terms in a formal letter that would follow the meeting. Spencer and her colleagues headed back to New York to present the Viners' requests to Twitter.

Spencer was cautiously optimistic. $1 million for nineteen or so creators in exchange for three posts a week might, she told me, "be the stop-gap we need to get us to the next place."

When the letter arrived a few days later, Spencer was aghast. The 1600 Vine group didn't want $1 million annually for the group; they wanted $1 million *each*, for a total of $19 million a year. Vine wasn't even profitable, and the company's standing with Twitter was shaky. She couldn't imagine how to pitch it: $19 million annually for a group of creators who posted juvenile and problematic content and were already top-fixing the app. It was not going to happen.

When Spencer told Johns the money wouldn't be coming, he was furious. Spencer tried to explain that Vine had no revenue, that the app was still being incubated by Twitter, and they simply didn't have the cash. Johns lashed out and accused her of lying. "I know what Twitter is making," Spencer said Johns told her. "They made $90 million last year." It's unclear where Johns got that number. While Twitter's annual revenue in 2015 was in the low billions, the company didn't turn a profit until the final quarter of 2017.

Spencer brought the deal to Twitter executives. They were against the deal, believing that paying creators could set a bad precedent. Years earlier, it had been a huge ask for celebrities to tweet for free. If Twitter started giving money to Viners now, then every celebrity would ask to be paid to tweet.

To make matters worse, news about the potential deal was already spreading among creators. Some 1600 Viners began bragging to Vine stars who weren't at the meeting that they were getting $1 million payouts. Pons told her good friend Gabbie Hanna, also a Vine star, about the details of the meeting. Hanna then called Spencer, livid, demanding to know why she was left out of the deal. "If you don't give me the same deal and pay me $1 million, I'm telling everyone on the app," she recalled Hanna saying.

As word of the potential deal spread, so did news of its failure. Spencer reported the bad news to the 1600 crew. After they got over the disappointment, they retaliated.

※

In the months that followed, the 1600 Viners all but ceased publishing original content to the app. Instead, they posted video after video encouraging fans to follow them on Instagram, Snapchat, YouTube, Musical.ly, and Facebook.

Still, Vine pursued a last-ditch effort to retain its talent. In February 2016, the company flew a handful of popular Viners to the company's London office. The plan was to help them network and collaborate with UK creators. Vine staff were going to put on a seminar about how to create the most engaging content. The company put the creators up in a posh SoHo hotel, while Gruger and others from Vine's U.S. office shepherded the Viners around town. The trip seemed to go well, and the creators were again posting original content to their Vine pages.

But the morning they were set to depart for London's Heathrow Airport, Gruger came down to the hotel lobby and saw the creators huddled together around their phones, chatting excitedly. Instagram had just launched a hub on its "Explore" tab with a little video carousel highlighting creators' Instagram profiles and videos. Every person featured in Instagram's creator highlight section was in the hotel lobby.

Instead of rising to the top, mid-level Viners, encouraged to defect by the 1600 Vine crowd, were fleeing the platform.

Staff left too. Toff departed the company in January 2016, and the app flailed without a leader. Others stepped into temporary leadership roles, but it felt to employees like a ship taking on water with too many holes to plug.

Hannah Donovan, a seasoned product executive who had worked at Last.fm and MTV, took over as the platform's new general manager in June 2016. Her mandate from Twitter CEO Jack Dorsey was to bring innovation back to Vine's creation tools. Vine's woes had bred deep resentment among staff and toward their own parent company, which they felt had fumbled Vine from the start. In early September 2016, Donovan was called into a seemingly mundane financial planning meeting for Twitter, the premise for which was to discuss the first-quarter budget for 2017. Midway through, someone mentioned that Donovan's head count at Vine would be frozen. Alarm bells went off in her head. Someone said, "Oh, she doesn't know yet." Donovan wasn't sure what was going on.

After the meeting she immediately reached out to Dorsey, who said they needed to talk. Back at her hotel room in San Francisco, she threw up. She knew what was about to happen.

Twitter as a whole was in a perilous spot throughout 2016, in part due to the broader industry's pivot to video. The app was hemorrhaging users, and its stock was cratering. Advertising revenue was flagging, and the media narrative was that the company was facing troubled times. "Twitter's growth has stagnated, and it's nowhere close to turning a profit," CNN declared. The company's product, engineering, media, and HR directors had all left the company by mid-2016.

Donovan knew all of this, so when Dorsey told her that Vine would be shuttering, she was shocked but not wholly surprised.

It fell on her to see Vine through to the end. With few options at her disposal, she made it her mission to preserve as much of the app as possible, fighting to archive web profiles so that creators wouldn't lose access to years of work.

On October 27, 2016, Vine announced it was shutting down. All but ten Vine employees were immediately laid off. Those who remained, including Donovan, spent their days archiving and arranging to relaunch Vine as a simple camera app. It would host all of Vine's existing videos, but no one could post anything new.

Vine was dead.

// CHAPTER 14 //

The Shuffle

Vine's demise shocked the online creator community. Never before had a massive social media platform shuttered overnight. Its collapse set off a scramble.

Porting an audience from one platform to another is terribly difficult. When Vine folded, a generation of small to mid-level creators halted. Many lobbied Vine to transfer their followers to Twitter accounts, but the company's internal tensions wouldn't allow it. "It was a dark, dark period," Karyn Spencer reflected. "We saw so many creators lose an audience of millions when the app shut down."

The death of Vine underscored just how important creators had become. Given their sheer scale, the 1600 Vine stars had an easier time pushing their audiences onto new platforms than smaller creators did. But the death of Vine ultimately meant the loss of the community and camaraderie that the app had created—both online and at the apartment complex.

By early 2017, when Vine officially closed, Rudy Mancuso and the other 1600 Viners had all but moved out of the apartment complex. The building's management had put up long enough with dozens of tenants running and screaming through the halls and filming pranks (like constructing a zip line over Hollywood Boulevard to send packages to fans, or pretending to be shot on the street as unsuspecting people outside looked on). Fans regularly crowded the building's entrances. The building had imposed restrictions, like banning filming by the pool and in common areas, but they were difficult to enforce. In March 2017, Logan Paul announced in a vlog that the building had chosen not to renew his lease. He was one of the last of that group to move out.

Everyone went their separate ways, online and off. Mancuso began associating more with film directors and built a career in traditional entertainment, eventually directing and starring in a short film for Amazon Studios about his life with synesthesia. Bachelor also ventured into traditional entertainment, starring in several Netflix movies and TV shows. But the majority of Vine creators who attempted to jump to traditional entertainment fell into the gulf. Now, without the distribution power of their Vine fan bases, they were just another name in Hollywood trying to score a break.

In Vine's wake, the dawning lesson might have been *not* to trust the fickleness of viral fame, that the traditional routes to success existed for a reason. But ultimately, the former Viners who pushed harder into the online world found the most success. The decisions that left Vine so exposed also fostered its lasting legacy by birthing an entire generation of online stars that, to this day, remain powerful figures on the internet.

Many top creators, like then twenty-two-year-old Alexandria Fitzpatrick, known online as AlliCattt, a Viner with 4.9 million followers, spread their content widely as Vine floundered. Eventually, Musical.ly became her primary platform, where she regularly live-streamed to her 2 million followers. "It's a much better, direct way to engage with my fans," she observed.

Gabbie Hanna was among those who successfully ported their audience to YouTube and Instagram before the outlook at Vine grew dire, but when the shutdown came, she found herself in mourning. "I complained a lot about it the last year or so but holy shit I'm sad it's over," she tweeted. "Vine being deleted is like an old friend I haven't really talked to in a while dying & even tho we aren't close anymore it hurts to see them go, but in all seriousness." At the time, Hanna had over 2 million followers on Instagram and YouTube.

Vine star Amanda Cerny was less nostalgic: Vine's expiration was "sort of sad," she told me, but "as far as the fans go, almost all of them translated to Instagram, YouTube, Snapchat and Facebook, which are bigger platforms for me now anyway. So it's not like I'm losing them." (Since she had about 4 million followers on Vine, she could afford to lose a few.)

Vine out of the way, rival tech companies aggressively courted top creators. YouTube and Instagram opened their pocketbooks, as did Snapchat, Musical.ly, and YouNow. After Cerny pivoted to posting on Snapchat, the company promoted her as one of the top ten most followed celebrities on the app. Many of the platforms were converging on similar models of shareable videos, which raised the stakes for recruiting celebrities from other platforms. There was one major factor that they were grappling with: the arrival of a new player, the biggest social media giant of all.

※

In the days following the infamous meeting at 1600 Vine, Jon Paul Piques felt dismayed. Like many Viners, he was strategizing about his next move. Then, he noticed his friend Lance Stewart gaining traction on Facebook.

"Bro, you have to hop on Facebook," Stewart told him in early 2016. "Do a compilation video of your Vines and I guarantee it will go viral," he urged. He knew from experience: Stewart had posted one such compilation at the end of 2015 that received over 100 million views.

Facebook was caught off guard by the video revolution that Vine unleashed in 2013. But it had no plans to stay behind. In 2014, when Zuckerberg addressed employees at Facebook's first community town hall, he underscored the company's focus on video. "In five years, most of [Facebook] will be video," Zuckerberg announced.

The platform had corporatized in the decade since its launch. When it threw money at a problem, it made waves. Within a year of the announced pivot, more than 500 million users were watching Facebook videos every day, with a total of 8 billion videos viewed every twenty-four hours, according to SocialMediaToday. Meanwhile, user-posted videos on Facebook increased by 94 percent in the United States in 2015, according to social analytics platform Hootsuite.

When Vine started to slip, Facebook doubled down. At Facebook's F8 developer conference in April 2016, Zuckerberg said that in ten years "video will look like as big of a shift in the way we all share and communicate as mobile has been." In the company's third-quarter 2016 earnings call, Zuckerberg again and again repeated that, like Snapchat, Vine and Musical.ly, Facebook was "video first."

One of Facebook's major priorities was to make a dent in the live-streaming market, via its Facebook Live feature. Zuckerberg saw live video as a natural bridge between social media and traditional video entertainment. "Live video, we think, represents an example of something new," Zuckerberg said in that same earnings call in 2016. "It's not the kind of traditional video experience. It's actually a more social experience."

Facebook's investment resulted in a significant increase in users going live in 2016, and it facilitated one of the most viral moments of the year. In April 2016, BuzzFeed live-streamed a group of staffers in the company's office wrapping an ever-growing number of rubber bands around a watermelon with the caption: "Watch us explode this watermelon one rubber band at a time!" "According to Facebook, more people tuned in at the same time to watch it live than any other video [on the platform]," BuzzFeed later reported. At its peak, the broadcast (which lasted forty-five minutes) had 807,000 viewers watching simultaneously, with over 5 million total views.

Soon, viral videos began originating on Facebook. Candace Payne, also known as "Chewbacca Mom," garnered over 180 million views of a video of herself trying on a Chewbacca face mask in her car. Such viral moments were unpredictable, which made them difficult for Facebook to plan around. Viral videos were often short clips shot on a phone. (Payne's video, taken in her car, clocked in at just over four minutes long.) The majority of media publishers, even digitally savvy ones, were still focused on written articles, so most video content that showed up in people's feeds was from friends' or amateur pages. Facebook's goal was to successfully eat into the tens of billions of dollars a year spent on television ads, as YouTube was starting to break into. To do so, the company needed a reliable slate of more professional, more engaging, more predictable content.

Facebook had established a presence at VidCon in 2016, offering an exclusive talent lounge for top creators where they could network with one another free of managers and agents. After VidCon, the company invited creators to tour Facebook's campus and meet with the teams there to give user feedback and discuss ways to work together on the platform's video efforts. It helped move things forward, but at more of a walk than a canter.

Then, Vine cratered, and Facebook saw a golden opportunity.

The person at Facebook in charge of recruiting these former Vine creators was Lauren Schnipper. Having joined the company in 2014 to help Facebook develop partnerships with digital creators across its ecosystem, Schnipper was already well known in the online creator world. She was especially known among YouTubers, as she'd previously served as head of production and development for YouTube star Shane Dawson, helping him attract millions of followers. As Dawson's producing partner and manager, she handled day-to-day production duties for his YouTube channel, and oversaw his expansion into TV, film, music, and podcasting projects—the sort of experience and credibility that made her ideal for the recruitment effort.

Schnipper began reaching out to Viners and making the pitch for Facebook. It looked like a perfect match: Viners had clear audiences, style, and professional content, and Facebook had enormous reach and a team at the company eager to work with them. In the months surrounding Vine's shutdown, almost all the large 1600 Viners established massive presences on Facebook. Along the way, they worked closely with Facebook representatives, who settled them into the app and shared posting strategies as well as support for issues with their pages.

Facebook also paid media companies and creators to produce videos and live broadcasts. The company inked 140 contracts with video creators totaling more than $50 million in 2016. As part of those efforts, Facebook began doling out as much as $220,000 for top YouTube and Vine creators like Andrew Bachelor, Lele Pons, Logan Paul, and others from 1600 Vine to post exclusive content on Facebook.

Piques was among the creators to sign on with Facebook. In April 2016 he agreed to a deal where the company would pay him up to $119,000 over the course of six months if he went Live on Facebook at least five times a month. His first video, a compilation of his best Vines, went viral and garnered over 150 million views on Facebook. Those were astronomical numbers, and his page rocketed up from 1,000 followers to over 500,000 in a single day. His content traveled so far in the News Feed that his page reached 7.5 million followers by the following year.

He wasn't alone. Nearly all the Vine creators saw immediate success. One video that Logan Paul posted at the time, where he did splits in famous locations around the world, amassed over 20 million views. Piques, Paul, Pons, and Bachelor leveraged Facebook to become royalty on the platform, with engagement regularly outpacing any media company or celebrity page.

Former Viners were optimistic that they could once again dominate a platform with their content and make real money while doing so. But while Piques and others were thrilled to receive their stipends, after their experience at Vine, they were anxious to hear about Facebook's long-term plans. While a hundred grand sounded great, it was significantly less than what a video with similar view counts could earn on YouTube through its ad-revenue sharing. Facebook said it was still figuring out its permanent revenue sharing model for video. The platform tested allowing some creators to place fifteen-second ads in videos, and some participants said that the pay was comparable to YouTube. But a year after Facebook had lured many of the 1600 Viners, a broad-scale monetization process had yet to materialize. On top of that, Facebook was pushing creators to publish longer and longer content, eventually mandating that videos be at least three minutes long in order to accommodate an ad at the beginning or in the middle.

"They didn't tell us a timeline for making money," said Piques. "They said they were working on it. I told my family I wanted to write a book called *I Got 2.8 Billion Views and Made $0*, because that's what happened."

As they waited, the creators also began to feel like Facebook didn't want anyone to make money, unless it was through Facebook. For instance, although sponsored content was taking off on Instagram and YouTube, creators believed sponsored content created on Facebook was down-ranked in the feed. Facebook representatives responded that they couldn't control the algorithm and that users ultimately decided what was worthy enough to share, which didn't placate the creators.

Facebook also appeared to cut off other sources of revenue for creators. After the 1600 Vine alums attempted to establish a similar reposting scheme on Facebook, where they'd reshare one another's content, sometimes in exchange for money, a Facebook representative implied to a creator that they would be punished for such agreements.

By the end of 2017, digital creators lost faith in Facebook. They still had no monetization options and felt used and discarded in a short-lived experiment by the massive platform. Facebook was thrilled to boost them and encourage them to post throughout 2016, which helped get Facebook's video products off the ground. But by 2017, larger media organizations moved in and were creating content at scale for Facebook. The energy on the platform had shifted. It was the first year of Trump's presidency, and hyper-politicized content now thrived, not goofy videos of former Vine stars rolling around L.A. on a hoverboard.

The 1600 Vine crew further splintered. Piques leapt deeper into Instagram, where short-form videos were still performing well, while Paul, Pons, and others concentrated on YouTube. Options that had seemed endless during the boom of 2015 were now starting to narrow again.

※

In the aftermath of Vine's fall, Snapchat seemed to be one of the most likely beneficiaries.

After Snapchat Stories had rolled out in late 2013, many Viners had gone back and forth between the two platforms. As Vine headed toward ruin, Snapchat exploded into mainstream culture and developed immense cultural capital.

Snapchat's golden era was led by the musician DJ Khaled, who used it to share daily updates with his fans. Khaled posted relentlessly, churning out a stream of videos of his adventures, often documenting his Jet-Ski rides around Miami. He coined catch phrases on Snapchat that became instant memes on the wider internet: "Another one," "Bless up," and "They don't want you to." No other social media app could approximate the real-time, free-wheeling, catch-it-while-you-can connection that Snapchat enabled.

What Kutcher had done for Twitter, Khaled did for Snapchat. "Occasionally there is a marriage of artist and medium so perfect it trumps everything that came before," the journalist Jon Caramanica declared in the *New York Times*. "No one has mastered Snapchat like. . . . DJ Khaled, who has become a social media celebrity in a way that outpaces his musical fame. . . . His effectiveness and addictiveness in the medium have elevated him from carnival barker to transcendent public figure."

DJ Khaled's success lured other celebrities and public figures, like reality stars Spencer Pratt and Kylie Jenner, onto the app. It also boosted Snapchat's homegrown talent, like doodler extraordinaire Shaun McBride. A host of Viners came over as well, including many from the 1600 Vine cohort. As the app garnered mainstream attention, many of the same companies that experimented on Vine turned to Snapchat for brand deals.

But Snapchat, too, ended up faltering in the aftermath. Ironically, it did so in much the same way that Vine did—by losing the trust of its creators. For existing Snapchat talent, the influx of Vine talent didn't look like a windfall; it looked like an invasion.

Snapchat and Vine had long had different features that gave Vine's stars greater reach than Snapchat's. Vine had a reshare feature ("revine"), while Snapchat did not. Vines were easily shareable across the internet, while Snapchat Stories vanished. All of these quirks meant that Vine's creators were able to amass larger fan bases than Snapchat's.

Snapchatters had worked for years to find creative ways to boost their audiences. One successful technique up until 2016 was to work with Viners on brand deals to increase their reach. "It was common for me to be on deals with Viners," said McBride. From 2014 on, he and fellow Snapchatters like Mike Platco paired up with Viners, including Logan Paul and Rudy Mancuso. Together, they could combine the Snapchatters' unique storytelling methods with the raw audience size that Viners offered. "Brands would showcase the story and artwork I did," McBride said, "then they'd be like, 'And we got this many views, which were because of the Viners.'" The Viners, for their part, treated it as a hedge against the fickleness of Vine, a way to diversify their audience.

By 2016, McBride and Platco had proven themselves capable of generating enormous engagement for Snapchat. But they felt like odd ducks on the platform. The app wasn't designed to help them grow their audiences or for average users to discover their content. "We thought, 'Is there anything that could help grow our brands or help us grow our content?'" recalled McBride. "But there never was, and there were never any features that could enable creators."

With their larger, preexisting audiences, the newly arrived Vine stars began gobbling up brand deals that would previously have gone to Snapchat stars. "We had a community of Snapchatters for a year," said McBride. "Then you had any Viner who had five or ten times the followers we had suddenly on Snapchat, so all the brand deals went to them."

Even as big Viners streamed onto Snapchat, gobbling up brand deals, the pool of money was shrinking. Instagram's version of Snapchat Stories meant that brands that previously put money into Snapchat campaigns were now splitting that money to advertise on Instagram as well as Snapchat. Snapchat's homegrown talent lost out on their home turf.

Snapchat soon found itself in a very familiar Vine spiral. "The disenchantment of Snapchat's top users calls to mind the downfall of Vine, which ignored its stars, who eventually decamped to YouTube and Instagram, leaving the platform a ghost town," BuzzFeed News reported at the time.

"Vine didn't know how to handle their creators, but Snapchat did not want creators," McBride opined. "I met with Evan Spiegel two to three different times and he was straight-up not interested in us. It was frustrating for us Snapchatters, because they just didn't want to have content creators. They didn't view Snap as a creator platform." As McBride saw it, the bosses just wanted Snapchat to be a messaging platform. "Chat" was in the name, and for Spiegel, that remained the game.

It was the rare founder who could look at the model he or she created and decide it needed to be chucked. And yet, if there was one thing that never changed about launching social media platforms it was that success required changing the program you'd launched. LinkedIn, which launched in 2003, followed the Facebook model. It functioned as a double opt-in network emphasizing real-world connections such as former colleagues, managers, and business associates. But the landscape changed, and soon it became clear that LinkedIn needed to pivot outside the friend bubble. In 2012, the platform launched its influencer program, which let a very small number of handpicked celebrities and business leaders like Bill Gates and Richard Branson write about trending topics.

As celebrities posted generic essays on management, a whole new class of creators bubbled up on the platform. By 2017, LinkedIn creators posting viral inspirational hustle porn known as "broetry" were gaining massive audiences, forcing the platform to adapt. It rolled out a suite of new tools making it easier for people to create and interact with content. In March 2021, the company finally introduced a "Creator Mode," which changed its Connect button to a Follow button, a one-way interaction that allows creators to build an audience more easily. LinkedIn also launched a podcast network, along with a $25 million investment in creators, and built a team of over fifty creator managers to help the platform liaise with its biggest influencers. This was the trend that Spiegel was missing: you could be a successful social network by empowering average users to become creators.

By 2017, Instagram had hired a team of partnerships managers to help usher creators onto their platform, and the company was highly responsive to top talent. Seeing how Instagram's creators were supported, Snapchat's biggest stars became fed up.

By the fall of 2017, Michael Platco, the Snapchat star, was aggressively moving his followers off Snapchat and onto Instagram. "I love Instagram," Platco told BuzzFeed News. "Every single bad thing I could possibly say about Snapchat, I could say the opposite of how my relationship is going with Instagram." And his frustration was borne out in the data. In October 2017, the marketing firm Mediakix found that top influencers who were active on both Snapchat and Instagram posted 33 percent less to Snapchat and 14 percent more to Instagram Stories throughout 2017.

That November, Snapchat dealt what felt like a final blow to its creator community, rolling out a highly controversial redesign aimed at separating "social" from "media" in an attempt to boost user engagement and differentiate itself from its competitors. The app was overhauled into two primary separate and distinct areas. Friends' content and Stories could still be found grouped together, but Snapchat stars and celebrities were now sequestered into dedicated channels on the app's "Discover" tab, which functioned as Snapchat's media portal. Suddenly, snaps posted by Kylie Jenner, DJ Khaled, and creators like Shaun McBride were sandwiched in among professionally programmed channels run by teams of people at media companies like Refinery29, *Vox*, and the *Daily Mail*. Snapchat stars' views plummeted.

Making it even worse, the app began pushing brands to run interstitial ads in the company's new Discover product, rather than working out deals directly with creators. It would have been hard to send a more direct signal to creators about how they were valued.

"I went a year before I said how dumb Snapchat was, because my entire life and career was built on Snapchat," said McBride. "I slowly and quietly fizzled out of Snap knowing they didn't like us. It was a slow death I saw coming. Finally I let it die." He shifted his focus over to a

YouTube channel he'd already started, which was on its way to gaining millions of subscribers. He also ran his own successful content company, Spacestation. But many Snapchat stars didn't achieve escape velocity. They continued to post to an ever-shrinking audience, before eventually giving up.

PART V

The Creator Boom

// CHAPTER 15 //

The Winners

When creators fled Vine, only three platforms were strong enough to catch them: Musical.ly, YouTube, and Instagram.

Musical.ly had been one of Vine's closest competitors. Vine's staffers dismissed it as an app for kids. That belief was rooted in truth, since Musical.ly was the first mainstream social platform that took hold primarily with Generation Z. But it was also an innovative app, whose robust video-editing tools gave users the ability to easily create polished, creative clips that would've taken hours to produce otherwise.

Musical.ly shrewdly positioned itself. For years, its bread-and-butter content was musical challenges, like the "Don't Judge Challenge" that went viral over the Fourth of July in 2015. The challenge consisted of making yourself look very ugly, usually using makeup, then pressing your hand up to the camera, pulling away, then revealing yourself looking beautiful, now with perfectly done makeup. The app encouraged large creators to participate in these challenges through in-app notifications.

The song associated with the challenge was a new single called "Cheerleader" by an artist named OMI. Both the song and the #DontJudgeChallenge hashtag became two of the biggest things on the internet as the challenge spread. Within days, "Cheerleader" shot to number 1 on *Billboard*'s Hot 100, and Musical.ly shot to the top of the app store charts. Off the back of that success, CEO Alex Hofmann positioned the app as a new force in the music industry, with coverage to that effect in mainstream outlets like MTV and the *Chicago Tribune*.

In the ensuing months, Musical.ly steadily chugged along, gaining traction. Top Musers like Ariel Martin, Blake Gray, and Jacob Sartorius joined DigiTour and monetized the same way other mid-2010s creators did: sponsored content. Then, in the summer of 2016, Musical.ly made a major push. Just a month prior to its first visit to VidCon, the app raised a fresh $100 million in investment, valuing the platform at $500 million. The company had good reason to be optimistic: when Vine floundered, Musical.ly, previously number two, became the number-one app in nineteen countries, including the United States.

For its VidCon debut, Musical.ly bought out a large booth in the convention center's hall. They set up a big couch on top of a stage, backdropped by a giant pink wall plastered with the Musical.ly logo. Stars from the Musical.ly stable, like Jacob Sartorius, performed every twenty to thirty minutes, drawing a crowd of over five thousand. The convention's organizers hastily coordinated with Musical.ly staff to make sure things didn't spin out of control.

Musical.ly also used VidCon 2016 as an opportunity to launch its own live-streaming app, Live.ly, designed specifically for Musical.ly users. Within just four days, Live.ly achieved over half a million installs and became the biggest live-streaming app in the United States. Live.ly adopted the YouNow model and allowed Musers to make money by receiving digital "tips" or gifts from fans. Some Musical.ly creators were suddenly making tens of thousands of dollars a week, giving them a much-needed stable revenue stream to grow their brands. "There were creators making $150,000-$200,000 a month through tips," said Hofmann.

Musical.ly's VidCon 2016 success cemented the company as a powerhouse. The platform was approaching 100 million downloads. Over 10 million new videos were being posted to the app daily. After seeing the attention Musers were getting at VidCon, big YouTubers and Instagram stars hopped on the app, but so, too, did digitally forward-thinking musicians like Ariana Grande, Selena Gomez, and Nicki Minaj. When Grande released her single "Into You" that year, she used Musical.ly to promote the song, which led to over 150,000 videos of fans lip-syncing and dancing along, boosting the song's streaming numbers and helping it go viral.

Warner Music Group became the first record label to sign a deal with Musical.ly in June 2016, allowing its music to be licensed for the platform. "Musical.ly has taken the art of lip-syncing, air guitar-ing and dancing with friends to a new level, making it one of the most compelling apps among young fans," a Warner Music spokesperson told *Billboard*.

In February 2017, Musical.ly experienced a major challenge from ByteDance, a Chinese tech conglomerate which had just acquired a start-up called Flipagram. ByteDance redesigned Flipagram to look exactly like Musical.ly. ByteDance also launched a Musical.ly clone within China called Douyin. While Musical.ly had its own tech and data science team, it struggled to replicate the powerful content-recommendation algorithm that ByteDance pioneered. China's population size also allowed Douyin to quickly capture more users domestically than Musical.ly had internationally.

Musical.ly management, feeling the pressure, focused on global expansion. But foreign apps weren't the only source of fear. Musical.ly's closest domestic competitors all had a tech behemoth behind them: Instagram had Facebook, YouTube had Google, and Twitch had Amazon. By mid-2017, Musical.ly was in active discussions for an acquisition.

After much discussion, Musical.ly's leadership decided the best strategic response to the Douyin challenge was "If you can't beat 'em, join 'em." That summer, they began meeting with ByteDance. In November 2017, the company sold to ByteDance for $860 million. ByteDance would leverage its command of artificial intelligence to improve Musical.ly's feed and help expand the platform's reach in Korea, Japan, and Southeast Asia—markets the company had previously struggled in. Musical.ly, in turn, would join ByteDance's family of apps. Executives drew up a list of four hundred possible names for the new app before landing on one they felt was exciting but neutral enough to convey the platform's broad ambitions: TikTok.

※

In August of 2018, Musical.ly relaunched as TikTok. Within two years it became the new social media powerhouse. Until then, however, the big winner was YouTube.

YouTube by 2018 looked like the elder statesman of social media. Well ahead of its rivals, the platform had understood that its creator class was a huge asset for the company and nurtured its talent accordingly.

A series of canny moves in the early 2010s had ensured that YouTube remained a hub for high-quality content. When YouTube realized it was missing the services offered to creators by MCNs, it acquired Next New Networks to fill the void. YouTube also offered a level of hands-on service to creators that boosted the overall quality of their videos. And throughout the decade, the company kept up its strategy of increasing its reach and reputation by investing directly in creators.

YouTube also repeated its successful program of offering grants to creators through cash advances, production help, and camera equipment. In 2011, YouTube gave $100 million to more than 100 new channels, including creators across a broad range of verticals, to supercharge their presence and create original, high-quality content for YouTube. In 2012, the company also introduced Creator Clubs, local meet-ups designed to help YouTube stars meet one another, collaborate, and share tips. YouTube Spaces launched in L.A., Tokyo, London, and New York, providing users with "free access to the latest and greatest equipment, sets and support to facilitate creativity and content innovation." YouTube expanded their offerings, allowing anyone to monetize their channel through ads, with creators taking a 55 percent cut of revenue to YouTube's 45 percent. YouTube continued to sponsor gatherings like VidCon and DigiTour, while also expanding its team dedicated to supporting creators.

YouTube took steps to help its creators monetize like traditional talent, a first among digital platforms. In 2014, YouTube announced Google Preferred, which positioned YouTube creators as "premium" content, worthy of blue-chip advertisements. The program was the realization of the far-sighted vision that George Strompolos articulated years earlier: home-grown creators could be as high-profile and in-demand as mainstream TV stars.

The lessons that YouTube learned through its Partner Program paid dividends. It knew for years that it needed to remain a place where a new generation of stars could emerge and thrive. But the YouTube platform had its own particular dynamics. Back in 2012, YouTube changed its algorithm to optimize for overall "watch time" rather than number of views, meaning that longer videos were more likely to be recommended. Leadership's embrace of watch time came out of its ambition to rival not only Facebook and other tech giants, but the entire medium of television. This seemingly small tweak changed the nature of content on the platform.

Before YouTube's algorithm change, users would hop frequently between short, catchy videos. Afterwards, creators noticed that the more they let fans into their daily lives, the deeper the engagement they'd generate on their videos. The algorithm was encouraging a return to the Lonelygirl15 era. It catalyzed a boom in vlogging that marked a kind of golden age for YouTube.

Casey Neistat, a New York City–based filmmaker, spearheaded the daily vlogging trend. After uploading videos on YouTube for years, he began posting daily vlogs on March 26, 2015, each eight minutes long. "What this vlog is for me," he said, "is a way for me to share my ideas and perspectives and have a conversation and chat in the comments and do Q&As. It's a forum for me."

His videos took off. Five months later, Neistat reached 1 million subscribers. He jumped to 4 million a year later.

While some of Neistat's vlogs brought in large numbers (his vlog of the Met Gala was a fan favorite), he hit a high point in January 2016. During a New York City blizzard, Neistat posted a video of himself, tethered to a Jeep, snowboarding the city streets. The video gained 6.5 million views on YouTube within twenty-four hours. That year, Neistat won British *GQ*'s "New Media Star" Man of the Year Award.

While a breathless, raw style fit Vine's hyperkinetic platform, YouTube's emphasis on total watch time demanded a different approach. Many Vine stars had already been producing daily content for Vine, leaving large stretches of footage on the cutting room floor when trimming to the six second max. Now they kept the camera rolling. To their surprise, not only did many of their fans enjoy a fresh format but they also appreciated having the chance to spend more time "together."

By the end of 2016, the Vine-famous brothers Jake and Logan Paul were posting daily vlogs to YouTube. Jake coined the catchphrase, "It's everyday [*sic*], bro," to speak to his relentless publishing schedule. David Dobrik, who had also hopped over from Vine, began uploading four-minute, twenty-second vlogs of his life at least three times a week. The former Viners were soon among YouTube's biggest stars, and the Partner Program rewarded them with ample ad revenue.

YouTube's algorithm didn't just reward watch time; it also rewarded frequent posting. A viewer was more likely to be recommended a given creator's video if they watched that creator regularly. Any creator who wasn't posting often could easily get crowded out by those who were.

In response, creators significantly accelerated their posting frequency. The production volumes created a cottage industry of video editors, production assistants, and other support roles. The creator industry entered a new phase, as stars began outsourcing production work and building entire teams to help run their content empires.

For many creators, especially the former Viners, the most natural way of keeping vlogs interesting was pranks. Pranks were easy to pull off, didn't require much budget, and were usually handsomely rewarded by the algorithm. When pranks failed to attract attention, top YouTubers engaged in public feuds and drama, many of which were staged. The YouTuber RiceGum became infamous for his diss tracks on other YouTubers, who often fired back with songs of their own.

Among the many vloggers who gained traction this way were Jake and Logan Paul. Jake, after founding his Team 10 content house, also engaged in public drama with members of other content houses at the time, including FaZe Clan and Clout Gang, which counted Paul's ex, Alissa Violet, as a member.

Paul released a song in 2017 titled "It's Everyday Bro," which amassed more than 200 million views on YouTube. The lyrics addressed his career as a daily vlogger and trashed Violet. In response, she and RiceGum, also a member of Clout Gang, put out a diss track titled "It's Every Night, Sis." The ensuing controversy boosted mainstream awareness of the content creators involved, a catty win-win.

Those who gained fame in the daily vlog era were loud, brash, and ostentatious. Soon, however, this behavior threatened their own success—and even the standing of YouTube itself. In November 2015, YouTuber Sam Pepper made headlines after uploading a video titled "KILLING BEST FRIEND PRANK." In the video, Pepper kidnaps a man and then forces him to watch his friend get "murdered" by a masked figure. It was essentially a staged snuff film, the victim crying in terror as he rocked back and forth in the chair he'd been tied to, helpless. The video went viral. Pepper was condemned by the media and even by fellow YouTubers, who felt like he took things too far. "YouTube please do something about sam pepper because i'm actually sick to my stomach and that should not be allowed on the internet," fellow YouTube star Issa Twaimz tweeted. All the attention only resulted in more views, and Pepper's follower count popped.

※

By the time Vine shuttered, over on Instagram the influencer economy was in overdrive. Instagram creators were regularly integrating sponsored products into their photos and videos, rarely with any disclosure. By 2016, brands were spending more than $255 million on influencer marketing every month just on Instagram, according to Captiv8, a company that connects influencers with brands.

As on YouTube, creators on Instagram invested in production as more money entered the space. While YouTube moved toward vlogs and juvenile pranks, Instagram favored a hyper-aspirational aesthetic. This aesthetic was defined by one color: Millennial pink.

"Wall scouting" was common among content creators in the mid-2010s, as they perpetually hunted for photogenic backdrops for their curated photoshoots. The bright pink wall at Paul Smith became the most iconic backdrop of this era. In 2016 alone, over 100,000 Instagram users uploaded a photo shot in front of it. The designer spent approximately $60,000 a year to service the wall, repainting it every three months. (Smith would claim that the wall was the second most photographed in the world behind the Great Wall of China.) Paul Smith even hired a security guard for the wall to ensure rules were followed. (No feet on the wall. No props, no costumes. No professional cameras allowed, though many skirted this rule.) Fittingly, in 2016, when Jake Paul announced his content collective, Team 10, the group did so in front of the Paul Smith pink wall.

The idea of an "Instagram wall" became so popular it transformed into a meme itself. Soon every event featured a pop-up wall for attendees to take photos in front of. So-called Instagram museums, like the Museum of Pizza and the Museum of Ice Cream, began appearing as pop-ups in cities across America in 2017 and 2018, luring influencers to their themed backdrops.

Bright, colorful walls; artfully arranged lattes and avocado toast; Millennial-pink everything, all with a carefully staged, color-corrected, glossy-looking aesthetic—this was the content that became synonymous with Instagram itself. No one capitalized on the Millennial aesthetic more than influencers. Some made thousands of dollars selling photo presets that would tint and color-correct anyone's pictures to fit this mold.

The Instagram influencer industry likewise birthed a cottage industry of photo and video editors, location scouts, and support teams. The concept of the "Instagram husband" emerged, signaling growing public awareness that producing online content often required a partner to manage all the behind-the-scenes labor such as photography and editing. (The phrase Instagram boyfriend or Instagram husband became shorthand for anyone who played that role, no matter their gender or relationship to the content creator.)

The concept of "Instagram husband" started in 2015. That year, a fake PSA produced by Jeff Houghton introduced the term to the masses in a viral YouTube video. With over 7 million views, the video profiled the men "behind every cute girl on Instagram." (They bemoan having to delete all the apps on their phone to make room for more photos and transforming into "a human selfie stick.") A Taco Bell ad released in the fall of 2018 parodied this concept. "I am an Instagram boyfriend," a man says while hanging off a carousel to get the perfect shot of his girlfriend. "Wing murals, candids, staged candids, I get them all."

For established creators, you had to spend money to make money: higher-quality posts cost more to produce, but those posts might attract upscale, well-paying sponsors. Lesser-known accounts also found opportunities as brands expanded their deals to include thousands of micro-influencers. There were plenty of ways to make it as new marketing dollars flooded in, each year's influx topping the last.

It seemed like the good times would last forever.

// **CHAPTER 16** //

Peak Instagram

THE ALARM BELLS RANG IN MARCH 2016, WHEN THE FEDERAL TRADE Commission went after Lord & Taylor. The company failed to disclose that it had paid fifty influencers up to $4,000 each to post photos to their personal Instagram accounts promoting the brand. None of those accounts made it clear that their posts were advertisements.

For decades, the FTC has required clear disclosure of advertising, but its guidelines have often lagged behind internet platform mechanics. Over the years, the FTC had issued a number of rulings, targeted at covert marketing and sponsored content on blogs, but social media remained a gray area. Instagram punished blatant ads early on for going against the spirit of the platform, but that was then. The whole point of sponsored content was to make it *not* look like an ad.

By the summer of 2016, industry watchdogs began calling out celebrities and content creators for deceptive advertising. *Bloomberg* reported that DJ Khaled failed to disclose that his Snapchat posts about Cîroc vodka were ads. Fashion lifestyle blogger Cara Loren Van Brocklin failed to disclose a sponsored post for PCA Skin sunscreen. Tavi Gevinson, who had come to fame when she started the fashion blog *Style Rookie* as an eleven-year-old, promoted her life in a luxury Brooklyn apartment building without revealing that she'd been given a discount on rent.

Members of the Kardashian family, who were famous for their sponsored-content deals, also came under fire. They followed tactics that bloggers had pioneered before them, weaving sponsored posts into their content and building up aspirational online personas. Even as most celebrities had yet to embrace the platform, members of the Kardashian family were using it to communicate directly with their fans, building a valuable and very monetizable bond with their audience. As Kim put it, social media was "my free focus group."

The Kardashians' investment in online spaces upended the format of modern celebrity. Kim launched a mobile app and a line of "Kimoji" custom emojis, netting her millions of dollars. (Her then-husband memorably rapped, "We made a million a minute.") Khloe launched a clothing line called Good American. Kylie launched a beauty and skincare line called Kylie Cosmetics. After Kylie copped to using lip fillers, requests for information about the injectables rose by 70 percent within twenty-four hours in the U.K. When she tweeted about not using Snapchat anymore in 2018, the company's market value plunged by $1.3 billion.

Back in 2016, a few months after the FTC confronted Lord & Taylor, Truth in Advertising, a consumer protection nonprofit, sent a "legal letter" to the Kardashians, lambasting them for their failure to disclose sponsored-content deals "as is required by federal law." Puma, Calvin Klein, Karl Lagerfeld, and Estée Lauder were among the twenty-seven companies the organization listed as part of the complaint. Truth in Advertising threatened to report the Kardashians to the FTC if the sponsored posts were not removed. Days later, recent posts from Kim, Kourtney, Kylie, and Khloe Kardashian were updated to include the hashtag #ad.

While this short hashtag at the bottom of a long post may seem slight, the issue was of tremendous importance in the online creator world. Some influencers' livelihoods depended on the money they made from unacknowledged endorsements. Creators worried that revealing which brands were paying them would kill their authenticity, repelling their audience. Even the influencers who strove for transparency were wary of angering advertisers by adding disclosure if they weren't explicitly asked to. Advertisers, for their part, thought consumers would be less likely to buy products if they saw influencers' posts as ads. Disclosure, they all understood, would stop the music.

By 2016, Instagram had stopped interfering with all but the most egregious promotional schemes on its platform. The company didn't tolerate outright scams, but it no longer wished to steer its fast-growing and influential user base away from monetizing. Plus, brand activity of any kind boosted Instagram's official ad rates and established Instagram as the place to be seen.

It fell on the FTC, then, to rein in the bonanza. In April 2017, it issued a warning to advertisers and content creators, sending over ninety letters to celebrities, athletes, and content creators as well as to the brands they worked with. Their message was clear: disclose sponsorships or face consequences.

The last thing advertisers wanted was for their multimillion-dollar marketing campaigns to lead to an embarrassing lawsuit. Nick Cicero, the former CEO of influencer marketing agency Delmondo, told *Bloomberg*, "For a lot of years it was really really loose, and you could get away with a lot more." Now, he was telling all of his clients to use the hashtag #ad without exception.

Fohr's James Nord said that after the FTC crackdown, the industry changed overnight. "If a post for a brand campaign that Fohr had organized didn't include disclosure, they could stop us from doing sponsored posts for two years," he said. "It would ruin our business. So all of the agencies and brands that worked with influencers took that very seriously. We started to push influencers to disclose. Influencers didn't want to disclose, but we had to make them do it."

Throughout 2017, more creators began to openly disclose sponsorships. YouTubers explained that they were being paid for unboxing videos. Instagram creators disclosed that the hotels they were staying in were comped. Musical.ly stars revealed that they were paid to promote the clothing brands they wore.

Instagram itself introduced a new feature designed to increase transparency around sponsored content and help users better differentiate between paid posts and organic ones. The feature displayed the phrase "Paid partnership with . . ." above sponsored posts and stories created by users promoting brands or products.

"There was a lot of fear," said Nord. Creators and advertisers alike worried that transparent disclosure would destroy everything that made the content creator space work. "People thought that this type of marketing was so effective because it didn't feel like an ad," said Nord, "and by saying it was an ad, people thought it wouldn't be effective."

Once the "paid partnership" labels and #ad hashtags appeared, everyone held their breath. They waited for engagement to plummet. Advertisers and creators alike readied themselves for big losses.

The media and many in the tech industry reveled in influencers' imminent downfall. Silicon Valley venture capitalists and journalists gloated that finally online creators would have get "real jobs." "[Influencers are] eager to assert the legitimacy of influencing as an actual business," one snarky story from 2017 read, "even though no one is exactly sure what influencing entails besides taking selfies."

But the crash never came.

"We all waited, but then nothing happened," said Nord. "It was a big nothing. No one's engagement dropped."

Advertisers and creators were shocked. Not only did their followers not seem to care that their posts were sponsored, some actually engaged with sponsored content at an even *higher* rate when labeled as such. "What we misjudged is that audiences actually thought it was cool that brands they knew and respected were working with creators that they respected and followed," said Nord.

The episode demonstrated that followers' admiration for their favorite social media stars was stronger than their distaste for ads. The bond between online creators and their audiences had become so deep that brand deals could be experienced vicariously too. Followers celebrated when creators landed bigger and bigger brand deals. Followers felt that they were playing a role in their favorite creators' success, which they were; when it came to brand interest, follower count and engagement metrics mattered.

Before the FTC ruling, most creators took on brand partnerships selectively. Once they realized their followers didn't care about ads, and in fact celebrated them, all bets were off.

"Once they became open with disclosures and the audiences didn't rebel, it was like floodgates opened," Nord said. "People's feeds went from five to ten percent sponsored content to forty percent sponsored content really quickly, because they felt like they had permission to do it." Now everyone wanted to know how they could cash in. "That's when the rat race really started in the creator space," Nord added. "Everyone wanted to know who was getting what deals, who was growing, who was not, and what were they charging."

The FTC could not have done creators a bigger service.

※

By 2017, brands had come to terms with the power of online creators. The right post by the right influencer could sell out a whole product line. Why not go a step further? Why not design the whole product line around the influencer?

Nordstrom was one of the first American brands that experienced firsthand the awesome power of influencers. U.K. fashion retailer Topshop had previously found success in collaboration with Kate Moss, inspiring Nordstrom to partner with street-style icon Caroline Issa on a new clothing line with prices ranging from $195 to $2,995. The following year, Nordstrom teamed up with Socialite Rank villain turned reality star turned street-style fixture Olivia Palermo. Together, they launched a yearlong clothing line available in select stores and online. By 2017, Nordstrom was ready to partner with influencers directly. Influencers would conceptualize and design the collections, while Nordstrom handled the back-end production and distribution. When the product line went live in stores and online, influencers would promote it to their followers.

At the time, brands that engaged in influencer marketing were simply happy to have an attractive person with tons of followers talking about their product. But Nordstrom was among marketers on the forefront, tapping into sales data and figuring out how to scale their partnerships.

"More than most companies, they're looking at what influencers actually convert into sales and why," Kate Edwards, co-founder of influencer marketing firm Heartbeat, told the website Glossy. "They're using that as a basis for deciding who to work with, as opposed to the way that most brands do, which is: Are they cute, and do they have a lot of followers?"

Arielle Charnas, age thirty, was no stranger to the influencer marketing universe. Having grown up in Long Island, she started a fashion blog in 2009 called Something Navy. She worked at the Theory store in the Meatpacking District and posted photos of herself in fashionable outfits, often featuring trendy garments and must-have accessories. She wrote in a voice that was elevated but accessible, and she became a fixture in the fashion-blogger landscape in the early 2010s.

Charnas was part of the wave of bloggers who invaded Instagram in 2015 and 2016, helping to shift the platform from being about photos *by* you to photos *of* you. Charnas had done advertising deals on her blog for years. In 2015, she signed a multi-year endorsement deal with TRESemmé and appeared in television ads for the haircare brand.

By early 2017, when Nordstrom approached her, Charnas was about to surpass 1 million followers on Instagram. She was already eager to launch a fashion line of her own. The two entered into an agreement where Charnas would design a line in collaboration with Treasure & Bond, a sub-brand owned by Nordstrom. She worked with the company to develop a thirty-piece clothing and accessory line, to be sold exclusively at the retailer. For months, Charnas teased the collaboration, leveraging Instagram Stories to show behind-the-scenes photos and videos, and engaged her followers in the design process by asking them to vote on patterns and colorways.

There was so much hype for the collection that when it launched in September 2017, Charnas's fans overwhelmed Nordstrom's website with orders. Nearly every item sold out within minutes. In less than twenty-four hours, her line generated over $1 million in sales. Fans took to her comment sections begging her to restock the items. Soon many of the pieces were listed on eBay for up to double the price. "The notion that she was able to move more than $1 million in product in less than 24 hours at Nordstrom is relatively unprecedented in the influencer world," *Women's Wear Daily* declared of the launch.

Charnas proved to brands and retailers that working with online creators could pay off in a big way. Until that campaign, most brands viewed influencers solely as a means of promotion. Buying ad space on a YouTuber or Instagram influencer's account had been an afterthought, a last step once a product was already designed and produced.

The Charnas line's success showed how effective it could be to bring online creators in at an earlier stage and grant them creative control. It wasn't just their reach or sense of style that made creators effective brand partners. Unlike, say, movie stars, creators like Charnas were also intimately familiar with the tastes and proclivities of their followers. They monopolized expertise about the products that appealed to their sizeable fan base. Creators gave retailers direct access to a dream market.

"As an influencer, my platform has given me the resources of real-life data and feedback by listening to my followers and seeing what they got excited about on my Instagram," Charnas said at a meet-and-greet at a Nordstrom store in Chicago following the launch. "We felt an obligation to use this incredible knowledge to give my followers what they wanted. . . . We are hoping to build upon this formula for success and continue to listen closely to my amazing audience."

By September 2017, 4 out of every 5 referrals to Nordstrom.com's mobile website were driven by an influencer, with RewardStyle, the affiliate marketing platform, accounting for 79 percent of those referrals. Nordstrom, eager to capitalize on the early success, ordered a restock of Charnas's entire line for December.

Eventually, Nordstrom gave Charnas an entire standalone brand named Something Navy, after her blog. With Something Navy, Charnas once again generated millions for the department store, and Nordstrom leaned harder into the strategy in later years. The company launched more content creator collaborations with influencers Chriselle Lim, Blair Eadie, and Julia Engel.

Nordstrom and other brands scaled their approach and began engaging with nearly every type of creator. Every piece of content contained a referral code, so that Nordstrom could track the creators' conversions and creators could further profit from the sales they drove. The company partnered with creators large and small, from Alexandrea Garza (736,000 YouTube subscribers) to Shea Whitney (71,000 YouTube subscribers). To advertise their yearly sale, Nordstrom offered Instagram stars referral codes and invited YouTube creators to vlog their shopping experiences. They paid makeup and lifestyle content creators and bloggers to post about their Nordstrom hauls on Instagram Stories and YouTube.

The company also began working with so-called micro-influencers (influencers with only a few thousand followers) and local influencers to launch custom lines in an effort to re-create what Charnas had done in key regional markets.

"A new summer collection from Nordstrom looks decidedly Texas-chic, which is no accident given its Dallas connection," a local Dallas news site declared in May of 2018. Cassie Freeman, a lifestyle creator with a fashion blog and tens of thousands of followers on Instagram, collaborated with Gibsonlook, a clothing line, on a limited-edition collection of women's clothing sold exclusively on Nordstrom.com.

Freeman's line sold well and opened her eyes to the type of deeper partnerships she could build with brands. "I remember the day Suzie [Turner, the head of Gibsonlook] emailed me about a possible collaboration," Freeman said. "I thought she meant an Instagram campaign or some small sponsorship. You can imagine my shock and excitement when I learned she actually meant an entire line of clothing based around my style."

※

As more influencers sought to productize themselves, management companies sprouted around them. Digital Brand Architects (DBA) was founded in 2010 and by midway through the decade it had become the premiere management company for lifestyle influencers. DBA's sister agency Digital Brand Products negotiated the most high-profile Nordstrom partnerships including those with Charnas, Julia Engel of Gal Meets Glam, and Rachel Parcell.

DBA was among the first to treat influencers as entrepreneurs. Over half a decade before the YouTuber MrBeast would be lauded by Silicon Valley men for mainstreaming the idea of creator-driven products, DBA and the women they represented were creating household brands. While the public and much of the media dismissed female influencers as vapid, DBA recognized them as savvy businesspeople. Some 92 percent of DBA's clients are women, and 94 percent of its team is female.

As DBA's initial crop of talent got married, bought houses, and had children, the company helped them spin out lines of cookware, home goods, and more through its product and licensing division. "From content creation and brand partnerships to product development and events, we have served as strategic advisors for creators . . . as they've evolved beyond the traditional monetization avenues and expanded into launching their own brands, investment ventures and other entrepreneurial pursuits," said Vanessa Flaherty, president of DBA. In 2018, DBA founded a podcast network called Dear Media in a joint venture with entrepreneur Michael Bosstick, leveraging podcasts as a springboard for female influencers.

"When we started DBA almost ten years ago, the creator economy didn't exist," said Raina Penchansky, DBA's chief executive officer. "Women are investing in other female-owned businesses; they're buying homes. Being able to help women grow in terms of their own businesses, it's part of our DNA."

In 2019, DBA's primacy in the field was cemented when the management company was acquired by United Talent Agency.

The boom in sponsored content in the late 2010s also had some strange consequences, consequences that brands could never have imagined even a few years earlier.

In 2018, a content creator named Palak Joshi posted a photo that looked like standard sponsored content: a shiny white box emblazoned with the red logo of the Chinese phone manufacturer OnePlus, shot from above on a concrete background. The post included the branded hashtag tied to the marketing campaign for the phone's launch, and her post tagged OnePlus's Instagram handle. The post was nearly identical to posts from the company and other content creators announcing the launch of OnePlus's new Android phone. Joshi's post, however, wasn't an ad. "It looked sponsored, but it's not," she told me. Her followers viewed it as just another sponsored post. "They just assume everything is sponsored when it really isn't," she said. Joshi said she wanted it that way.

Sponsorship was no longer seen as selling out. By 2018, a brand deal had become a status symbol. If you got a good brand deal, you were seen as successful; if not, you were negligible. Content creators began faking sponsored content. Sydney Pugh, a lifestyle influencer in Los Angeles, staged a fake ad for a local café, purchasing her own mug of coffee, photographing it, and writing the caption in a highly promotional way as if to sound like it was sponsored. "Instead of [captioning] 'I need coffee to get through the day,' mine will say 'I love Alfred's coffee because of A, B, C,'" Pugh said. "You see the same things over and over on actual sponsored posts, so it becomes really easy to emulate, even if you're not getting paid."

Creators paid their own way for luxury vacations but posted as if the airlines and hotels had comped them. They tagged the clothing they wore and promoted the food they ate as if it were given to them for free.

When Christian Dior relaunched its famous Saddle bag in a massive marketing blitz in mid-2018, scores of creators were gifted the $2,000-plus bag and paid to post about it on social media. The campaign was chaos—and a perfect encapsulation of the state of influencer marketing. While some creators who were actually paid to promote the bag failed to disclose they were part of the campaign, other creators who were never part of the campaign simply purchased their own bag and posted as if they had been invited to participate in the campaign, tagging their posts with #ad.

While the FTC had been explicit that paid advertisements needed to be disclosed, there was no rule against pretending that a post was a paid advertisement. Monica Ahanonu, an illustrator and Instagram creator, told me at the time that fake sponsored content was so rampant that she was losing track of what was sponsored and what wasn't.

Members of Generation Z, who were coming of age immersed in this ecosystem, were some of the most likely to fake sponsored content. "People pretend to have brand deals to seem cool," one fifteen-year-old told me. "It's a thing, like, I got this for free while all you losers are paying."

"We did a big NARS campaign and there were a bunch of people posting #ad that weren't in the campaign," said James Nord. "We asked them to take it down. We said, 'You're not part of this.' But it became this validation."

"In the influencer world, it's street cred," said Brian Phanthao, a young lifestyle influencer. "The more sponsors you have, the more credibility you have. It's really common with kids in high school. They're very influenced by influencers."

Most marketers were thrilled with the free advertising, but high-end brands were frustrated. They worried that their brands were being devalued by teenagers and low-level content creators acting like they'd partnered on campaigns. Some resolved this by creating lists of "official partners" or highlighting the creators they had real partnerships with on their own Instagram feed. Few users consulted these references, however, and the imposters campaigned on.

※

A 2018 survey conducted by the World Federation of Advertisers found that 65 percent of global advertisers planned to increase spending on influencer marketing in the next twelve months. But as more and more aspiring creators wanted to cash in on the sponsored-content gold rush, the industry's infrastructure began to buckle.

In response to growing demand, hundreds of influencer management and marketing services emerged to act as middlemen. More than 420 such firms launched in 2017 alone. By the following year that number nearly doubled.

These new companies nearly all functioned the same way. A content creator would sync their social profiles with the influencer marketing platform. The platform would in turn pull in their follower counts and other analytics offered by social media APIs. Users entered basic demographic information, indicated their topics of coverage (such as lifestyle, food, or beauty), and listed their rates.

Brands could use these databases to sort thousands of creators by criteria including age, follower count, geography, and rates. Influencer marketing and management platforms were most beneficial to smaller and mid-level creators. Stars with 10 million followers or more usually had dedicated managers and agents sourcing and negotiating deals for them. But for the rest, influencer management and marketing platforms were an easy way to source and streamline deals.

The platforms made money by taking fees or charging brands a certain percentage to run the campaigns. For instance, a brand like Coca-Cola might pay an influencer marketing platform $100,000 to run a campaign to reach young women between the ages of twenty-five and forty. The influencer marketing platform would then assemble a roster of influencers who could reach that targeted demographic, coordinate their posting to ensure creators used the correct marketing copy from the brand, and distribute the funds once the campaign was complete—after taking a cut, of course.

Even with all the money being made, content creators were still getting stiffed. Because they were now sourcing deals through middlemen, influencers were at the mercy of the go-betweens' tactics. Some marketing brokers failed to pay the content creators at all, or delayed payments as much as six months or over a year after campaigns.

The wheel had turned again. Given the increased competition, creators were being asked to pay up-front costs to participate in branded campaigns. For instance, they now had to book their own flights to participate in a brand campaign, then apply for reimbursement with the company.

Across the influencer marketplace, gaps emerged and operators squeezed in. It became cheaper and more attractive for marketers to outsource their responsibilities to various sub-agencies, which increased the size and importance of the secondary market for creator services. Things were moving so quickly that standards and regulations never had an opportunity to settle. Sometimes influencers who felt they'd struck gold on a brand deal ended up penniless. In those cases, no matter how many image filters you applied, the only color you were left with was blue.

// CHAPTER 17 //

The Adpocalypse

As the influencer industry on Instagram flourished, YouTube was about to deal a massive blow to its creator ecosystem.

On Valentine's Day 2017, the *Wall Street Journal* published a story that rocked the internet: "Disney Severs Ties with YouTube Star PewDiePie After Anti-Semitic Posts." The piece offered graphic descriptions of several videos posted by YouTube's most subscribed creator, Felix Kjellberg, a twenty-seven-year-old Swede known online as "PewDiePie."

By 2017, PewDiePie had amassed over 53 million YouTube subscribers. He was doing multimillion-dollar deals and was signed by Disney-owned Maker Studios, the premier MCN. The first season of his YouTube-backed reality show, *Scare PewDiePie*, was a massive success. Kjellberg belonged to YouTube's Preferred program, which gave him and other elite YouTubers exclusive access to lucrative advertising campaigns, among other perks.

Kjellberg attained all of this by posting a steady stream of comedy skits, gaming, and commentary videos. He was steeped in meme culture and internet ephemera. Many of his videos appealed to what his fans called "edgy humor."

When exposed to the public, however, his jokes were seen in a different light. The *Wall Street Journal* reported on nine recent videos that included anti-Semitic jokes or Nazi imagery, specifically one video posted on January 11, 2017, where he paid two men in India to hold a sign emblazoned with the phrase "Death to all Jews."

The video was the latest trend in YouTube's escalatory prank cycle. In this version, creators engineered stunts by paying random, and often unknowing, accomplices through the website Fiverr, which helps people find gig work.

Many fans saw the stunt as social commentary on the extreme things people, especially in developing nations, would do online for money. They were bolstered in this interpretation by Kjellberg's claim that he didn't think those he paid would go through with the task. Outsiders, however, found the prank horrifying.

When the *Journal* article came out, Kjellberg faced unprecedented backlash. Maker Studios promptly cut ties with him. YouTube canceled the second season of his reality show. The Indian men in the video apologized, saying that they didn't speak fluent English and didn't understand what the message meant.

Kjellberg expressed frustration with the controversy, which felt to him like a bad-faith attack. In a video defending himself, he said, "If I made a video saying, 'Hey guys, PewDiePie here. Death to all Jews, I want you to say after me: Death to all Jews. And, you know, Hitler was right. I really opened my eyes to white power. And I think it is time we did something about this.' That is how they're essentially reporting this, as if that's what I was saying."

Months after the PewDiePie story went viral, Michael and Heather Martin, of the YouTube channel DaddyOFive, were publicly lambasted for subjecting their children to terrors on camera for views. Onetime family vloggers, they'd moved on to pranks that included pouring disappearing ink on their carpet and blaming their son Cody for it. "What the fuck did you do?" yells Heather to summon Cody to his room. "I swear to God I didn't do that," cries Cody as his parents berate him. Cody's face turns red, and he crumples to the ground.

The Martins attempted to clear their name by claiming their videos were scripted and everyone was acting. But viewers on the internet didn't buy it. Thousands of Reddit users reported the family to Child Protection Services in Maryland, where they lived. The Martins ceased creating content temporarily as a result of court-ordered probation and temporarily lost custody of their children. Their sons have since returned to YouTube, operating a more benign channel called "The Martin Family."

Such backlash did little to break the prank cycle, as journalist Amelia Tait observed in her writing on YouTube's prank culture in 2017. "Boyfriends pretend to throw their girlfriend's cats out windows; fathers pretend to mothers that their sons have died. YouTubers deliberately step on strangers' feet in order to provoke fights. Sometimes, yes, pranksters are arrested for faking robberies, but in the meantime their subscribers continue to grow in their millions." The algorithm had spoken.

※

Years had passed since YouTube launched its first effort to sell premium advertising against its creators' videos. The company had succeeded beyond its wildest dreams. Advertisers were more than happy to pay out the nose to reach top creators. Those creators, in turn, were rewarded with dollars and subscribers in the millions. For the younger generation, figures like PewDiePie were bigger than movie stars. But now YouTube's top stars were proving that they could be a liability. Many had spent years posting inflammatory content with impunity.

YouTube's algorithms had catapulted PewDiePie and other controversial creators to the top of the charts. If the crassness and exploitation on YouTube was highly visible, its societal impact would ultimately be minimal in comparison to the content amplified by algorithms at Facebook and Twitter.

In the aftermath of the 2016 election, after years of positive coverage, reporters now scrutinized how Facebook and Twitter influenced political discourse in the run-up to the election. Democrats, searching for reasons behind Donald Trump's stunning victory, grew increasingly furious as evidence emerged about Facebook's role. The company seemed to have incentivized fake news; granted access to the personal data of unsuspecting users to questionable political consultancies; and constructed an algorithm that prioritized engagement over everything else, promoting polarizing content and amplifying bad actors. Twitter's own algorithms did much the same. Throughout Trump's campaign, his own Twitter account proved to be his most powerful asset.

Trump was joined by a band of alt-right trolls and content farmers like Ben Shapiro, Mike Cernovich, Milo Yiannopoulos, and Lauren Southern, who had all become political social media influencers. These users boosted the MAGA message through social media posts, Facebook and YouTube videos, and live-streams, while monetizing in the same way that more vanilla creators and influencers did.

As Twitter and Facebook became the major focus of post-election investigation, controversies on YouTube were viewed through a political lens. The content of some of PewDiePie's videos was not far out of sync with some of the alt-right trolling that had been influential on Twitter. While Kjellberg eventually pleaded that his video was taken out of its prank-era context, YouTube itself was drawn deeper into controversy.

Alongside the *Wall Street Journal* story, other reports emerged of disturbing activity at YouTube. Whistleblowers from inside YouTube called out the company's recommendation engine for serving up ever more extreme content and rewarding conspiracy-laden content. The London *Times* discovered through an investigation that advertisements for hundreds of large, multinational companies including Mercedes-Benz, HSBC, and L'Oréal were airing on YouTube videos created by terrorist groups such as the Islamic State and Combat 18, a violent pro-Nazi faction.

The media storm prompted scores of companies to reevaluate their YouTube advertising strategy. Large brands had long known there was risk in running ads on YouTube given the lack of editorial oversight, but they'd been willing to take that risk as YouTube seemed to operate on its own island: the chance of backlash seemed low, and the platform allowed them to reach the younger audiences they couldn't find elsewhere. But now YouTube and its creators were no longer low-profile. They were in the spotlight, for all the wrong reasons.

Advertisers pulled out en masse, and YouTube began hemorrhaging money. In March 2017, AT&T and Johnson & Johnson, two of the largest advertisers in the United States, announced they would stop their ads from running on YouTube and other Google properties. They cited the platform's inability to guarantee that the ads wouldn't appear next to hate speech or offensive content. Enterprise, the car-rental company, and Verizon quickly followed. Numerous British advertisers also pulled out, urging YouTube to make drastic changes to the way it served ads.

Phil Smith, director general of Isba, an organization that represented hundreds of British advertisers, told the *Guardian*, "Whatever Google's editorial policy, advertising should only be sold against content that is safe for brands. . . . Google should ensure that content is quarantined until properly categorized."

The stakes couldn't have been higher for Google. The year prior, it made $19.5 billion in net profit, primarily due to its successful advertising business. Philipp Schindler, Google's chief business officer, published a blog post that sought to quell advertiser concerns.

The post began as a defense of the creator ecosystem. "The web has opened a door for new communities and platforms that help people find diverse views and have a voice," Schindler wrote. "Today, anyone with a smartphone can be a content creator. . . . Google has enabled millions of content creators and publishers to be heard, find an audience, earn a living, or even build a business. Much of this is made possible through advertising. Thousands of sites are added every day to our ad network, and more than 400 hours of video are uploaded to YouTube every minute. We have a responsibility to protect this vibrant, creative world—from emerging creators to established publishers—even when we don't always agree with the views being expressed."

Schindler then acknowledged the company also had a responsibility to its advertisers. He apologized for the ads that had run next to offensive and hateful content, and he promised a full review of the company's advertising policies and tools. The company also said it would roll out changes that gave brands more control over what sort of content their ads appeared next to.

As the company rolled out tighter restrictions, they had dramatic ripple effects. The change, which became known in the online creator community as the "Adpocalypse," affected nearly every notable creator. Immediately, top YouTubers saw their earnings from ads plummet. Certain videos and channels were demonetized, often without clear reason. For creators who'd spent years learning YouTube's rhythms and algorithms, it was disorienting. They'd come to depend on the platform to maintain their livelihoods, and now the bottom was dropping out.

Kjellberg was among the most vocal creators to speak out, saying that his revenue had cratered following the controversy. The damage wasn't confined to PewDiePie and his ilk. Creators across every corner of the internet panicked over lost revenue. The Internet Creators Guild, a nonprofit organization representing people who make content online and run by YouTuber Hank Green, surveyed dozens of creators and found the losses in earnings were significant enough to wipe out many creators' entire business. "My ad revenue is WAY down from 2016 and 2015. Less than half. So depressing," one creator wrote below the Creators Guild's video.

Ethan and Hila Klein, whose commentary channel h3h3productions is among the most well-known of its kind on YouTube, claimed that by May 2017 they were making just 15 percent of their pre-boycott earnings. Philip DeFranco, who ran a popular YouTube commentary and news channel, said that his ad earnings dropped by 80 percent following the advertiser boycott. Within a couple months they rebounded slightly, but he was still 30 percent down by mid-April. Progressive news commentator David Pakman said his ad earnings dropped 99 percent.

Creators scrambled and formed group chats to share information about what was happening. The changes YouTube was rolling out were opaque to them, and it wasn't clear how they could avoid running afoul of certain filters. For instance, YouTube now allowed advertisers to opt out of videos that contain "tragedy and conflict" or "sensitive social issues," as well as videos that were "sexually suggestive." Algorithms scanned creators' content, and if anything in their videos fell into these broad categories, their videos were effectively demonetized. Did that mean a college lecture on *Macbeth* was a no-go? How about a trailer for the new Star Wars movie? If someone made a video about how to make chocolate chip cookies, could someone object because many cocoa growers in Africa are exploited? It was impossible to know what would clear the bar.

"New options for advertisers that is resulting in most of us not having ads," Ethan Klein tweeted in April 2017, including a screenshot of YouTube's new ads policy. "The best part is that @TeamYouTube doesn't explain this roll out or how to protect yourself, they just silently screw everyone as usual."

As the company made its changes, it tried to placate both advertisers and creators. "We know this has been a frustrating time and we'll continue to update you as quickly as we can as things evolve," a YouTube staffer named Marissa wrote on a company forum to worried creators seeking help. The company representative also urged creators to reevaluate their video thumbnails, titles, and descriptions to make sure that they were advertiser friendly. What exactly that meant, however, was still unclear.

※

If the spring 2017 Adpocalypse had been all, YouTube might have stabilized the ship. But in early November, another wave broke. A *New York Times* investigation on children's content creators on YouTube found that kids were being exposed to animated videos containing characters engaging in suicide and violence. Days later, the writer and artist James Bridle, in a blog post titled "Something is wrong on the internet," detailed additional content intended for kids on YouTube that could only be described as horrific.

What Bridle and the *Times* found was an array of videos, many with tens of millions of views, that were bizarre and downright abusive. One video, uploaded to YouTube by a verified creator named Freak Family, showed a young girl crying as someone shaves her face seemingly as punishment. Blood appears to spill from her forehead. It was given the

SEO-friendly title: "Bad baby with tantrum and crying for lollipops little babies learn colors finger family . . ." and it had amassed a stunning 53 million views. This and other disturbing videos seemed to have been engineered to exploit YouTube's recommendation algorithm in order to queue up after a video that parents had chosen for their children. YouTube's algorithm was ushering children down a nightmare-generating rabbit-hole.

"I don't even have kids and right now I just want to burn the whole thing down," Bridle wrote. "Someone or something or some combination of people and things is using YouTube to systematically frighten, traumatise, and abuse children, automatically and at scale, and it forces me to question my own beliefs about the internet, at every level."

What Bridle found became known as "Elsagate," named after the protagonist of the hit Disney movie *Frozen*. Bridle didn't mince words: "To expose children to this content is abuse. . . . What we're talking about is very young children, effectively from birth, being deliberately targeted with content which will traumatize and disturb them, via networks which are extremely vulnerable to exactly this form of abuse. It's not about trolls, but about a kind of violence inherent in the combination of digital systems and capitalist incentives. It's down to that level of the metal. This, I think, is my point: The system is complicit in the abuse."

Once again, YouTube scrambled to contain the controversy. It seemed as if many of the bizarre kids' videos had been created at scale, perhaps using AI, and crammed with keywords to appeal to YouTube's algorithm. A YouTube spokesperson was dispatched to the media to reassure worried parents that these videos were rare. In fact, they were a symptom of YouTube's decision to take an AI-first approach to moderation. The site was so big now that it was a vast ecosystem—impossible to be directly observed in full, and even harder to police. YouTube had long had a dedicated Kids portal, with different rules about acceptable content, but clearly the wrong videos were sneaking through the cracks.

To combat the crisis, YouTube doubled down on human moderation of kids' videos, and it tightened up its algorithms even further.

YouTube CEO Susan Wojcicki even considered removing ads from children's videos altogether, which would have been an unprecedented change. Ultimately, the company ended up removing ads on more than 2 million videos that targeted children and cracked down even harder on views that its algorithms deemed dicey.

It was a major step, but advertisers were still skeptical of the platform's ability to police itself. Once again, they began pulling back their YouTube ad spending and cutting off creator sponsorships.

Top creators vented about it in their YouTube videos. Phil DeFranco, Hank Green, PewDiePie, and Casey Neistat all urged YouTube to get it together. Their ability to make a living as content creators was at stake.

"We've talked about the Adpocalypse in the past and we've seen different forms," Phil DeFranco said in a video. "There's a ton of frustration because YouTube's trying to figure something out, but in the meantime people are just losing money. So many YouTubers are working paycheck to paycheck. When you see the Jake Pauls of the world, it's easy to go, 'Oh, YouTubers are all rich.' Please know that's not the case. There are a ton of creators that are just barely making it."

Then, Logan Paul vlogged a dead body.

※

Shortly after Christmas, Logan Paul and friends traveled to Japan. They explored the country all week, gathering material for his vlog. At the time, Paul posted a daily fifteen-minute vlog of his life to his more than 15 million subscribers on YouTube, or the "Logang" as he called them.

As part of their Japan tour, Paul and his friends decided to pay a visit to the Aokigahara forest by Mt. Fuji, known as the "suicide forest." They said they wanted to camp out and "look for ghosts." But Paul was also playing with the thrill of potential controversy. The forest got its name from the hundreds of people who took their own lives there, and Paul and his friends ventured in with their cameras rolling. The result: on December 31, Paul uploaded his daily vlog, a fifteen-minute video with the title, "We found a dead body in the Japanese Suicide Forest. . . ."

"This is not clickbait. This is the most real vlog I've ever posted to this channel," Paul says in the beginning of the video. "I think this definitely marks a moment in YouTube history, because I'm pretty sure this has never hopefully happened to anyone on YouTube ever. Now with that said: Buckle the fuck up, because you're never gonna see a video like this again!"

Cameras in hand, Paul and his buddies stumble upon a dead body in the forest, seemingly just hours after the man took his own life. A hundred yards into the brush, the group freezes and Paul points at a tree. A dead man can be seen hanging with a noose around his neck. "This isn't a fucking joke, guys. . . . Call the police, bro," he says to the camera.

The vloggers then approached the body and filmed close-up shots of the corpse hanging, with his face blurred. The camera zooms in on the man's hands, and Paul says, "Dude, his hands are purple. He did this this morning."

"I'm so sorry about this, Logang," Paul continues. "This was supposed to be a fun vlog." The group heads to the parking lot to regroup after the incident. Paul pulls out a bottle of liquor from his backpack and quips, "I'm already getting flagged for demonetization, bro, fuck it," then chugs the bottle.

At the end of the video, after issuing a PSA-style warning about suicide, Paul openly ponders whether asking for subscribers would be crass. "Is it bad if I do the subscribe thing? Maybe," he says. "Nah, man, this is why I do it. So people can live the journey with me. If you're not a part of the Logang, make sure to subscribe. Tomorrow's vlog—I promise you—will be much happier."

The video was outrageous enough to have caused controversy even under normal circumstances. But in the aftermath of the PewDiePie controversy and Elsagate, the mainstream media was readier than ever to condemn offensive YouTube content.

The backlash to Paul's video spread across the internet. He tried to apologize. "I didn't do it for the views. I get views. I did it because I thought I could make a positive ripple on the internet. . . . I intended to raise awareness for suicide and suicide prevention," he tweeted at the time. But his apology didn't line up with the tenor of the video, and Paul couldn't quell the backlash.

"A YouTuber . . . claiming it's just a fun joke hyping up his friend who just showed someone dead for views is the exact summary of YouTube in 2017," tweeted Nate Garner, a social-media influencer with nearly 3 million followers.

"Dear Logan Paul, How dare you! You disgust me. I can't believe that so many young people look up to you," tweeted actor Aaron Paul (no relation). "So sad. Hopefully this latest video woke them up. You are pure trash. Plain and simple. Suicide is not a joke. Go rot in hell."

"I make one video about my eating disorder and my entire channel is demonetized forever but Logan Paul can show a dead body and make fun of suicide and go . . . on trending," wrote another YouTuber.

For over a week, the Paul controversy was inescapable. It was discussed on CNN and TV morning shows across the country and covered by every mainstream media outlet.

YouTube itself did damage control. "We expect more of the creators who build their community on YouTube," the company said in a statement. As they had done with PewDiePie, Toy Freaks, and others, they demonetized Paul's videos and booted him out of the Preferred advertiser network. "We are looking at further consequences," the company said.

YouTube's liaison to top creators, Graham Bennett, later told *Bloomberg* tech journalist Mark Bergen that the string of controversies was a wake-up call. "It was the first time we realized that YouTube creators were legit global stars," Bennett said. "And that meant that if they did something out of line or crazy and newsworthy, it will be news everywhere in the world." The comment revealed that even YouTube itself didn't recognize how significantly the media environment had shifted over the decade prior, and how much influence online creators actually wielded.

In the aftermath, YouTube developed a broader code of conduct for creators. A new phrase began making the rounds internally: "Making money off YouTube was a privilege, not a right."

※

Three strikes weren't enough to count YouTube out, however. With the alarms still blaring, the company threw its whole weight behind appeasing advertisers. To try to suppress extremist content, abusive videos, and the bizarre content, it made over three hundred tweaks to its algorithm in 2017. At the same time, the company tried to wrangle its talent and hold them to a higher standard.

Creators took a different lesson from the Adpocalypse of 2017. YouTube had been the most supportive platform to them in the 2010s. Now, online creators were reckoning with just how dependent they were on YouTube. It wasn't just the algorithm or the platform; the Adpocalypse also drove home the fickleness of advertisers as a whole. Moreover, with the new level of scrutiny, many creators felt like they were one bad post away from losing their livelihood.

They began to diversify into other sources of income. Large YouTubers set up pages on Patreon, a service that allows creators to earn subscription revenue in exchange for exclusive content and other perks. Mid-level creators followed, especially after YouTube rolled out more restrictions around monetization for small YouTube channels in January 2018. It's hard to upsell on Patreon, however, and many parents of young kids who make up the majority of top influencers' audiences were wary of signing up for any sort of recurring subscription.

Many YouTubers then took a page out of the Instagram influencer playbook. Some leading creators had sold merchandise for years, but as the Adpocalypse progressed, merch emerged as a dominant business model for the platform's creators. Almost no one embraced the pivot to merch harder than the Paul brothers, and their success paved the way for a generation of smaller creators to follow. From 2016 into 2017, more than forty top YouTubers, including Casey Neistat and Jiffpom, a dog with more than 7 million Instagram followers, launched their own custom merch lines. When Paul's channel was demonetized in January of 2017, he leaned even harder into his merchandise line, "Maverick." Soon, big YouTubers like Paul, David Dobrik, and the Nelk Boys were constantly shilling branded merchandise in YouTube videos, promoting limited edition "drops" of hoodies, sweatshirts, T-shirts, and more, raking in millions of dollars in sales.

Online creators who launched merchandise lines in the early 2010s usually did so with the help of their MCN or print-on-demand ecommerce storefronts. By 2017, however, a start-up called Fanjoy began to dominate the merch market. Fanjoy was founded in 2014 by then twenty-six-year-old Chris Vaccarino as a company that initially created custom branded care packages for fans of bands and musicians. It was like Birchbox, but for music fans.

Then, in 2016, Vaccarino met Jake Paul. Vaccarino noticed more and more young people getting famous online, and while they ran advertisements regularly, they didn't seem to have any sort of robust merchandise strategy. "I noticed these kids who were getting hundreds of thousands of likes on Instagram," Vaccarino told the *Daily Beast*. "So, I just went to kids' Instagram pages, looked at their YouTube pages, and emailed them."

Jake Paul was one of the first to respond. At the end of 2016, the two launched Paul's first sweatshirt line. Vaccarino quickly realized that, unlike traditional artists and celebrities, online creators like Paul were viral marketing masterminds. Promoting yourself wasn't stigmatized in the influencer world the way it was in traditional entertainment.

"These kids are super young and hungry," Vaccarino said of his primarily eighteen-to-twenty-year-old clients. "YouTube stars usually haven't done merchandise deals in the past, so the way they approach selling their products is radically different than older, more traditional stars."

Creators like the Paul brothers were relentless when it came to plugging their own products. On Jake Paul's 2017 Christmas album many of the lyrics were devoted to selling merch. The album featured songs such as "Fanjoy to the World," which included the lyric: "*Fanjoy to the world, my merch has come.*"

"When he read me the lyrics, I was like, 'Dude, are you serious?'" Vaccarino said. "I mean, who else has done that? Other people would be scared to post something like that song but with Jake, he's like, 'What else I can do to go over the top and get people talking?'"

The chorus of another Paul holiday song, titled "All I Want for Christmas," is "*Buy dat merch. Buy dat merch. Buy dat merch. Buy dat merch. Buy dat merch. Buy dat merch. Buy dat merch.*" Paul then proceeds to spell out the URL of his merchandise website on Fanjoy. The Christmas album was released during prime holiday shopping season in 2017, and the merch drop reportedly generated tens of millions of dollars. A *New York Magazine* investigation in 2018 found that in Jake and Logan Paul's most recent fifty videos, the brothers urged their viewers to buy their merchandise at least 195 times.

The Paul brothers expertly orchestrated a steady stream of viral stunts to promote their products. Jake rented out a billboard on Melrose Avenue in Los Angeles with a giant picture of his brother Logan's face on it with words that read "I LOVE THIS MERCH!!! BUY SOME!" and a link to Jake's Fanjoy shop. Logan pranked a fellow content creator known as KSI by coordinating a hand delivery of his Maverick merchandise to his home in London before the two were set to compete in a boxing match. Fanjoy shipped close to 1 million pieces of merch in 2017. Throughout 2018, growth continued as creators sought to diversify their revenue streams away from ads.

"YouTube is turning into a giant merchandise-plugging factory," *New York Magazine* reported in 2018. "Old vlogger standbys like confessionals, stunts, and pranks are rapidly and shamelessly being shoved out in favor of endless plugs for merch. And why not? It's become almost impossible to make money on YouTube otherwise."

In early 2018, PewDiePie joined the party, releasing a new unisex merch line called Tsuki. Around the same time, police were dispatched to the streets of NYC after 10,000 David Dobrik fans swarmed his "Clickbait" merch pop up with Fanjoy in SoHo.

The YouTube superstars rode the merch wave, and their young followers gobbled up the pitch. But not every creator could sell millions in merch, and even those who could still found themselves buckling under the demands of daily publishing schedules. The Adpocalypse was a shock to the system, and its after-effects ricocheted around YouTube and the rest of social media for years. Online content had appeared to creators to be a welcome alternative to the stresses of a day job. But it was increasingly starting to feel like its own torturous rat race.

// **CHAPTER 18** //

Breakdown and Burnout

IN EARLY 2018, ELLE MILLS, A NINETEEN-YEAR-OLD FROM OTTAWA, Canada, began making a name for herself on the internet by embracing the prank scene—but with her own spin. Her pranks weren't necessarily wholesome, but they were a far cry from the problematic content that often dominated young people's feeds.

Her first viral stunt, posted in March 2017, featured her interviewing her friend's Tinder dates, whose answers were hilarious. She followed it up with antics like throwing herself a parade, staging herself a funeral, stealing her brother's identity, and surprising her parents with new tattoos. She quickly garnered over 1 million subscribers on YouTube and raked in over 75 million views.

Mills was able to walk the tightrope between chasing views and growing her brand responsibly. It was a tricky balancing act. "The thing about prank culture on YouTube is that it's so driven by numbers and views," Mills told me. "It's something where you can understand why people do stuff that gets them in trouble." The month prior, a YouTuber creator named Arya Mosallah had filmed a series of fake "acid attack" prank videos—amidst a slew of real acid attacks in the U.K.—drawing a sea of backlash.

Mills was intent on keeping her own content lighthearted and fun. At one point, she flew to Las Vegas and legally married her sister's boyfriend for a vlog. The video generated over 2.7 million views but caused a massive headache when she struggled to end the joke marriage with a real divorce. "I find that sometimes I feel like I almost crossed the line for views. I come up with ideas like, 'This is insane! No one has ever done this before!'" she said. But, on the whole, she added, "I'm lucky to have family and friends who say, 'Hey, that crosses the line. You shouldn't do that.'"

Mills continued to make a name for herself, but she found that it required a punishing schedule. From Tuesday through Sunday, she scoured the internet and drafted video concepts. Her content required enormous effort and planning, and she poured every ounce of her energy into making videos. Mills procured intricate props and staged her shots meticulously. Then she'd shoot the video, which could take hours or days. She'd spend all weekend editing. Maybe she took a break on Monday—maybe not.

Even by early 2018, she said that her videos caused her an "unhealthy amount of stress." In one video titled "Dear Viewer," Mills said into the camera, "The fact [is] that I think everything I make isn't good enough, and the fact that I cry every week because it's a never-ending cycle. The best way I can describe what I'm going through is like having to go through kindergarten through college in one night and being expected to get straight A's and not let anyone down."

In addition to YouTube, Mills maintained an active presence on Instagram. She also launched a branded merch line featuring T-shirts and hoodies with her logo: a Canadian maple leaf, a soda bottle, and a camera.

As Mills worked hard to raise her profile, she began to get mainstream press and interest from sponsors. That year, NPR highlighted Mills's work and called it a mix of "Lena Dunham in *Girls* and a Woody Allen film." New opportunities came rolling in.

Mills already felt like she was barely treading water, and then came additional pressure to seize the moment. She couldn't say no to the opportunities she'd worked so hard to get. Mills set off to travel, hosting meet-and-greets with fans. She set up brand deals with Fortune 500 companies like Wendy's and Samsung. She kept up with her grueling content schedule, as well as with the social media expectations that came with it, engaging directly with fans and fellow creators.

On top of all this, Mills was also reckoning with another rising trend on YouTube and other platforms: the growth of cringe, drama, and tea (gossip) channels, which treated creators and fandoms as characters in one vast soap opera, all pitted against one another.

※

It's no surprise that gossip became the lifeblood of the creator universe. Now that YouTube stars had tens of millions of followers and resembled mainstream stars of the past, there was a vast audience who wanted to know everything about their lives. Fans giddily speculated on their romances, their next career moves, and their controversies.

Daniel Keem, a YouTube institution known online as Keemstar, founded one of the original drama channels in 2014 called DramaAlert, dubbed the "TMZ of YouTube." The channel began as a way of updating the internet about beef between online gaming streamers, but then it expanded and added staff as it struck a nerve and provided daily coverage of top influencers.

Other drama outlets soon followed. Instagram page Diet Prada emerged to cover the celeb and fashion industry, Comments by Celebs revealed personal interactions and storylines between Hollywood stars, and Deuxmoi emerged to share anonymous celebrity news tips. Thousands of tea accounts across Instagram and YouTube, and later TikTok, posted updates about content creators in every genre. The Blogsnark and GOMI communities shed their forums and formed a sprawling network of "snark" subreddits, where users trade gossip on specific influencers or types of creators. "Over the past five years," the journalist Rebecca Jennings reported for *Vox* in 2020, "Instagram accounts, Snap channels, TikTok pages, and profitable media companies have been built around which YouTuber is dating which Twitch streamer, who's breaking up with who, and who's possibly getting cheated on. . . . In a world where there are far too many famous people for traditional tabloids to keep track of."

One media outlet called Famous Birthdays built an entire business cataloging internet stars. In 2012, Evan Britton, the site's founder, saw young people's interest in online creators rising and recognized an opportunity to build a mobile-first celebrity encyclopedia. Wikipedia still relied on traditional metrics of fame, and the mainstream media didn't often cover influencers unless there was a scandal. "I realized there was a big gap between who the industry thought was famous and who really had the fame," Britton told me. The site took off and by the latter half of the 2010s it had surpassed *Entertainment Weekly* in web traffic and had four times the readership of *Teen Vogue*.

By 2018, the online drama industry had become a major draw that could drive not just views but entire commentary accounts to prominence. That year, a feud erupted between several of the biggest beauty YouTube stars on the planet. Jeffree Star, a former MySpace star turned beauty YouTube king, had long been close friends with fellow YouTube makeup stars Manny Gutierrez (aka Manny MUA), Laura Lee, and Nikita Dragun. But soon the dynamic foursome grew apart. After a series of controversies, Gutierrez, Lee, and Dragun stopped hanging out and collaborating with Star.

The three became closer with another beauty YouTuber named Gabriel Zamora. Everything came to a head in August 2018 when Zamora tweeted a photo of himself with Dragun, Gutierrez, and Lee, giving the camera the middle finger along with a comment believed to be directed at Star. "Bitch is bitter because without him we're doing better," the caption read.

The image and its caption set off a tsunami of gossip across the internet that became known as "Dramageddon." Drama channels tirelessly documented and analyzed every update related to the friendship's implosion. They dug up old tweets, monitored creators' likes and comments, tracked their subscriber numbers, and cataloged fan reactions.

"YouTube channels that cover drama on the platform saw huge growth thanks to their coverage of Dramageddon, as fans were desperate to learn more about the feud," *Insider* reported. "Drama YouTuber Dustin Dailey, for instance, had many more videos go viral after meticulously covering the drama." Viewers of drama channels seemed to have a bloodlust for building up and tearing down influencers, and the real-time subscriber counts rewarded that behavior.

This new breed of gossip journalists focused on the internet formed an emergent industry that would grow to overshadow traditional entertainment media in the coming years. With the breadth and immediacy of their coverage, they effectively rendered legacy tabloids like *US Weekly* and *Star* magazine irrelevant. But it didn't take a veteran reporter or a larger organization to capitalize on creator gossip.

Some creators realized that they could massively increase their followings simply by creating content that bashed or picked fights with YouTube megastars. Cody Kolodziejzyk (known online as Cody Ko) was a millennial former Vine star who had shifted his efforts over to YouTube after Vine shut down. By 2018, he realized that he could garner a wide audience making videos reacting to "cringe" content from well-known creators. He would pull up videos of his internet peers and tear them apart, often making fun of their jokes and skits.

Cody Ko attacked the Paul brothers and other prank YouTubers. The bigger the star he targeted, the more views his videos racked up. He also attacked the former Vine star Lele Pons, who now had an active YouTube and Instagram presence, mocking her videos and criticizing her for making content ostensibly for children but with overt sexual references.

The majority of Kolodziejzyk's videos were rightfully calling out problematic behavior by influencers. He lambasted "views culture," where creators were expected to do anything for attention and views online, claiming it was destroying people—and he had a clear point during the prank era. To some of the creators who were targeted by Cody Ko, however, he was taking their work out of context and sending a wave of harassment their way.

A group of four content creators known as the Dobre brothers, all in their early twenties, called out Kolodziejzyk for "cyber bullying" and challenged him to a boxing match. Jake Paul eventually confronted Kolodziejzyk in a video for his own vlog called "confronting internet bully cody ko," which was ironic, given Paul's own long history of driving bullying and harassment.

The worst harassment launched by fans of Kolodziejzyk's videos targeted female creators like Lele Pons. Misogyny was a constant of online life—dating back to mommy bloggers, Julia Allison, and the MySpace Scene Queens—but women on YouTube were held to impossible standards from which their male peers were exempt. "Cody Ko made it cool to hate," said Lele Pons. "He is the reason why I get hate." She said that after he began critiquing her, she developed severe obsessive-compulsive disorder. "With his videos, mentally I got more tics, more OCD, more mood swings," Pons said.

※

Harassment and bullying had been a serious problem on social media platforms since the MySpace era, but tech companies had never prioritized addressing it, allowing misogyny and hate to spread with abandon. Popular female creators were often forced to deal with a barrage of rape and death threats. As drama channels stoked outrage, many well-known YouTubers were brought to a breaking point.

Elle Mills was sent straight into the storm. Female YouTubers were largely expected to stick to creating "girl" content about beauty, fashion, and lifestyle. Pranks, stunts, and comedy, which Mills was creating, were seen as men's territory. Successful female creators like Liza Koshy, Lilly Singh, Colleen Ballinger, and Mills worked to shatter that glass ceiling, but each had to pay a steep price.

As Mills's profile rose, more and more people online attacked her and scrutinized her every move.

Mills realized she needed to take a break and reset from the constant abuse, but she didn't see *how* she could do so. She and other creators were working around the clock. Every day they hustled to piece together content in an increasingly competitive field. They juggled a punishing work schedule, navigated the whims of flighty brands and platforms, endured the stress of performing in the public eye, and dodged scammers all while opportunistic drama channels mined their content for the tiniest slip-up. "If you're not actively creating, or if you're going on a trip and you haven't actively created content to publish during said trip, you will go effectively back to the back of the line," Jon Brence, the director of talent at Fullscreen told NPR. Being an online content creator didn't come with paid vacation time.

Taking a break from posting on YouTube, Instagram, or livestreaming meant more than just a temporary drop in viewership and revenue. Taking a break sent a very negative signal to the algorithms that now rewarded frequent posts and repeat viewers. If a creator fell out of that cycle, they could see their livelihood vanish.

Aspiring creators like Mills had been grinding nonstop. Most established star creators kept up near daily posting habits. But by the middle of 2018, the camel couldn't bear another straw.

Top content creators began breaking down. First it was a handful, but then, emboldened, more said they were done with the hours, the pressure, and the abuse. Many quit their careers as online content creators, while others planned to step away and take time off to reassess their careers.

Mills was one of the first to kick off the wave. "This is all I ever wanted," Mills said of her career as an online creator in a 2018 video titled "Burnt Out At 19." "So why the fuck am I so fucking unhappy?" She explained that her anxiety and depression were getting worse and worse, to the point that she was having regular panic attacks. In the video, she Googled herself and showed how she'd been portrayed in headlines, and how YouTube commentator channels spoke about her when she tried to step back. She included footage of her crying and shouting alone under the covers in a hotel bed, on the phone with a friend who pleaded with her that things would get better. She ended the video explaining that she would be taking indefinite time away to focus on her mental health.

Mills was far from the only creator feeling this way. Casey Neistat released a video titled "The Pressure of being a YouTuber," addressing the issues creators were facing. "I've often discussed the pressures of being a YouTuber and it's a tricky thing to talk about because to find success on YouTube is to live the dream," he said. "This is the ultimate. And if you achieve this kind of success on this platform, which so many people try to do, like, how dare you complain about it? It is difficult to talk about because unless you've been in this position, I think it's challenging to empathize with it."

If Neistat was feeling it, likely everyone was feeling it too. Once YouTubers started talking about it, they realized they'd all been stuck on the same hamster wheel.

In a seminal piece on the wave of content creators quitting their careers titled "YouTube's Top Creators Are Burning Out and Breaking Down En Masse," journalist Julia Alexander captured why creators felt trapped: "Constant changes to the platform's algorithm, unhealthy obsessions with remaining relevant in a rapidly growing field and social media pressures are making it almost impossible for top creators to continue creating at the pace both the platform and audience want."

The golden era at YouTube faded to black. The company had fostered creators for a decade. It saw the potential early and developed a visionary plan of sharing significant revenue with creators. It did so before there *was* even any revenue, before "creator" was a word people used. Now that being a creator had not only become a potential livelihood but a *dream* career to many, YouTube had let creating on its platform become indistinguishable from every other terrible job.

✳

Since all the big creators maintained accounts across platforms, what happened at one platform inevitably spread to the other. After years of the Instagram aesthetic growing more and more polished, something finally broke. The gap between influencers' super-professional posts and the everyday lives of most users grew too large. "The influencer" tumbled from alluring idol into self-parody.

The first obvious sign of trouble was the Fyre Festival. In late 2016 and early 2017, over four hundred high-profile influencers and celebrities dubbed "Fyre Starters," including Kendall Jenner, Bella Hadid, and Hailey Baldwin Bieber, began promoting a festival that was being sold as "the cultural experience of the decade," with tickets costing between $1,500 and $250,000.

In March 2017, hundreds of celebrities and content creators simultaneously posted a single bright-orange square to their Instagram feeds. Users were directed to a video of stars (again including Bella Hadid and Hailey Baldwin Bieber) dancing and partying on an island in the Bahamas. The captions all contained some version of "So excited to announce the first ever #fyrefestival @fyrefestival fyrefestival.com."

Thousands of people purchased expensive tickets to the festival, only to find out too late that they had been conned. The promised fancy meals turned out to be cheese sandwiches. The luxury accommodations turned out to be FEMA tents. There was a severe shortage of bathrooms. None of the promised musical acts showed up. The festival was canceled, but only after attendees had arrived on the island. Every influencer involved with it faced a backlash for promoting what many called a scam.

Many influencers who promoted the festival, it turned out, were paid thousands of dollars to do so. (In Kendall Jenner's case, $250,000.) Fans and followers who'd previously ingested sponsored content began to question whether their favorite content creators were simply posting about things for money without a care in the world for their followers.

This was the backlash to sponsored content that every marketer had expected once ads were properly disclosed; it simply took several years to materialize. Part of the turn-off was simply that the Instagram aesthetic had aged like milk. According to Taylor Cohen, a former digital strategist at advertising agency DDB, the Instagram aesthetic's saturation point came somewhere around mid- to late 2018. "It's not the same as it was even a year ago," she told me for a 2019 article in the *Atlantic* titled, "The Instagram Aesthetic Is Over."

The transition could be seen through the ill-fated expansion of the Happy Place, an Instagram museum that opened to great fanfare in

Los Angeles in 2017. The Happy Place billed itself as the "most Instagrammable pop-up in America." When it opened, people were thrilled to fork over the nearly $30 admission price ($199 for a VIP pass). But when it arrived in Boston in mid-2019, it landed with a thud.

The same period saw a shift from polish to a certain kind of self-consciousness over what Instagram had become. "Instagram vs. reality" photos boomed in popularity in 2019 as creators attempted to make themselves seem more accessible. At Beautycon, a beauty festival in New York in 2019, Instagram stars spoke about moving away from ring lights and toward showing off their faces in sunlight.

As the public became more aware of the prevalence of sponsored posts, beauty influencers abandoned branded shots for ones that showed off their "empties" (empty bottles of product they *actually* use). Many accounts garnered hundreds of thousands of followers by calling out the various cosmetic procedures celebrities and influencers have undergone. Influencers also started actively speaking out about burnout, mental health, and the stress that comes with maintaining perfection.

"Everyone is trying to be more authentic," said Lexie Carbone, a content marketer at Later Media, a social media marketing firm. "People are writing longer captions.... I think it all goes back to, you don't want to see a girl standing in front of a wall that you've seen thousands of times. We need something new."

That something new was driven by a younger generation. While Millennial influencers had hauled DSLR cameras to the beach and mastered complicated photo editing software to get the perfect shot, by 2019, Generation Z was happy to post directly from their phones. In fact, many teens began going out of their way to *lower* the quality of their photos. Huji Cam, which makes your images look like they were taken with an old-school throwaway camera, was downloaded over 16 million times by 2018. Adding grain to your photos became such a big thing that Instagram began incorporating filters into its Stories feature to make the quality of photos look worse.

"For my generation, people are more willing to be who they are and not make up a fake identity," Reese Blutstein, a Gen-Z creator who amassed more than 238,000 followers in just over a year by posting unfiltered, low-production photos of herself in quirky outfits, told me at the time. Anything that felt staged was as undesirable for Blutstein's cohort as unfiltered or unflattering photos would be for older influencers.

The changing trends reflected evolving user preferences, but they were also a reaction against the burnout-inducing standard of the peak influencer era. "We all know the jig is up," said Matt Klein, a cultural strategist. "We've all participated in those staged photos. We all know the stress and anxiety it takes. And we can see through it. Culture is a pendulum and the pendulum is swaying. That's not to say everyone is going to stop posting perfect photos. But the energy is shifting."

For the reigning influencers, the shift was disorienting and even catastrophic. "What worked for people before doesn't work anymore," said James Nord. In 2018, a creator could post a shot with manicured hands on a coffee cup and rake in the likes. By 2019, people would unfollow. According to Fohr, by the end of that year, 60 percent of influencers in his network with more than 100,000 followers were losing followers month over month. "It's pretty staggering," he said. "If you're an influencer [in 2019] who is still standing in front of Instagram walls, it's hard."

"People are just looking for things they can relate to," said one Instagram user, and "the pink wall and avocado toast are just not what people are stopping at anymore."

As on YouTube, the trends that fueled the explosion of the influencer industry on Instagram were now leading to a fracture. The platform itself by late 2019 was morphing into a messy content amalgam of stories, "IGTV," photos, and video clips. For many users, photo posts simply became a vehicle to vent in the captions or comments section.

✸

Neither YouTube nor Instagram was dying. Instead, a massive shift in the creator world was underway.

If there was anyone who embodied this shift, it was Emma

Chamberlain. In 2015, Chamberlain was an average fourteen-year-old high school girl, a cheerleader and gymnast in San Francisco. That was when she began experimenting with making videos. She filmed herself and friends using her webcam and concocted dance routines that she'd edit down for Instagram. In 2017, she posted her first video to YouTube. Less than two years later, she had become one of the fastest growing YouTubers on the planet and revolutionized the way creators approach the platform.

While other top creators were burning out, Chamberlain was blazing her own trail. By 2019, she surpassed 7 million subscribers on Instagram and 8 million subscribers on YouTube. When she launched a podcast, it promptly hit number one on charts in fifty countries. She was heralded as "YouTube's it girl" and "the undisputed queen of teenage YouTube," with an engagement rate that would make any YouTuber weep. "Her growth has been so fucking insane," said MaiLinh Nguyen, a producer and YouTube strategist who has worked for several top YouTubers.

Becoming an overnight success on YouTube almost never happened anymore by the late 2010s. Several generations of YouTubers rose and fell since the platform was founded in 2005, but by the time Chamberlain launched her channel in 2017, it was largely accepted that the only way to make it big on YouTube as a new creator was by relentlessly posting and collaborating with larger channels.

Chamberlain didn't log onto YouTube with big aspirations. Like many girls her age, she simply treated the platform as another place to express herself. Her early videos were mostly fashion-related, with titles like "City Inspired Summer Lookbook 2017" and "How to: DIY Rose Patch Vans." She vlogged about taking her driving test and why she needed a fidget spinner.

On July 27, 2017, she uploaded a video that transformed the internet: a dollar store "haul" showing off all the products she bought at the store. "I made that video going to the dollar store. That was apparently a YouTube trend at the time, and it ended up working in my favor," she told *W* magazine. It was a very searchable topic, and her video ended up amassing half a million views, while she'd previously been averaging about 1,000 views per video.

After that post, her channel caught on like wildfire. Her videos suddenly garnered hundreds of thousands of views each. Teen girls in particular couldn't get enough. Chamberlain's videos were like nothing they had ever seen before. While most influencers relied on hyper-produced, staged videos with bright thumbnails and clickbait all-caps titles, Chamberlain posted lo-fi vlogs of her life as a high schooler, using default fonts, clashing color schemes, and lowercase titles that only underpromised. She often looked like she just rolled out of bed. She notoriously shunned makeup and sometimes skipped a shower. She didn't care if she looked weird or posed the camera at an unflattering angle. In one video she filmed herself in the car with the phone on her lap, pointing up at her face while she drove.

What brought Chamberlain's videos together, however, was an expert editing style that enhanced her humor. She was able to blend her personality with an editing style that made a twenty-minute-long video feel like a quick clip.

Chamberlain became famous for adding facial distortions and zoom effects similar to the zoom feature on Instagram Stories. She'd focus in, slow down, or add crazy music and sound effects to enhance what was happening in the video. Throughout her vlogs she interspersed her own reaction shots of herself while she was editing, giving voice to what the audience might be thinking as they watched. "It's a very memey editing style," said Nguyen.

"Emma was one of the first people to take that aesthetic of social media being very polished and all these influencers having these videographers and photographers follow them around all day and flip it," said Nguyen. Nothing about Chamberlain's videos looked "premium."

The shift away from hyper-polished content had already been happening on Instagram. But Chamberlain is the one who popularized lo-fi content on YouTube. Her ascendancy—occurring right when many top creators were reaching their breaking points—was very much a reaction to traditional influencer-style content on YouTube. Chamberlain lived in L.A. now and was undoubtedly a multimillionaire, but the point was she didn't take herself too seriously. She and her cohort were leading a movement against the aesthetic that had come to dominate both YouTube and Instagram. In many ways it was a return to what had made YouTube compelling in its early years.

Chamberlain's style gave birth to a genre that people called either "slacker YouTube" or "relatable YouTube." While she became synonymous with teen YouTube by 2019, she wasn't the only one growing quickly by being "relatable." Summer McKeen, then twenty, began vlogging at age thirteen and by 2019 had grown to 2.3 million subscribers on YouTube. "I don't feel like I need to put on a different face or put on an act when I'm filming, I just feel like I'm hanging out with my friends," said McKeen. "My camera is my friend. It's always been that way."

The catch, however, is that it still took an enormous amount of work to make a video look casual. Chamberlain spent twenty to thirty hours editing each video. "I've cried multiple times after posting a video," she told *W*. "It's like giving birth. Like, 'Oh, my God, that's my masterpiece.' And every single video is like that for me. So much work goes into each video that I don't know how I'm still alive." McKeen also spent hours on each video. "It's the same amount of work as the old way," she said. "It's just a more casual feel to the video." But rather than spending those hours making sure she looked perfect, the hours were spent ensuring the video effectively communicated her sense of humor and personality, which felt like a relief to many female YouTubers.

Even so, the very name that the genre had taken—"relatable YouTube"—was a misnomer. Chamberlain and her cohort were relatable to many, but not to all. The creator universe was historically almost all white at the top, and even as Chamberlain and her cohort bucked one dominant aesthetic, the movement remained one color.

In a video called "The Relatable White Girl Trend" a young Black vlogger who goes by AsToldByKenya on YouTube skewered the style. "This is the criteria to become a relatable white girl," Kenya continues. "You have to be pretty, you have to be funny, you have to curse but not curse too much that you're seen as vulgar, but you still have to curse to be seen as relatable. You have to have cute outfits. . . . You have to be a teenager. If you're not a teenager, you can't fall into this category, I'm sorry. . . . You have to be attractive to boys, but not actually come across like you're attractive to boys."

In another video titled "I Was A Relatable Youtuber For A Day! *cough* Emma Chamberlain," Don Bbw, a young Black YouTuber, says, "I actually don't know what's relatable, because what's relatable to you, might not be relatable to me," then jokes that he "forgot the most relatable part of being a relatable YouTuber, having a pimple."

"I feel like that [Emma Chamberlain] aesthetic trend is definitely a suburban white girl trend," Abby Adesanya, a talent manager, told me at the time, adding that the look is something YouTubers of color mostly don't have access to. "These YouTubers of color, their whole vibe is received so differently than the Summers and the Hannahs and the Emmas." In the winner-take-all environment of social media, that meant real sums of money making their way to white creators while YouTubers of color watched. A perfect example involved the phrase "on fleek," which appeared in an Ariana Grande video and a Denny's commercial. Kayla Lewis, the sixteen-year-old Black girl who introduced the phrase in a Vine, wasn't paid a penny. It was hard to avoid the conclusion that had it not originated with a young Black girl, the outcome might have been different.

PART VI

Influence Everywhere

// CHAPTER 19 //

TikTok Dominates

On August 2, 2018, Musical.ly relaunched as TikTok in over one hundred countries around the world. To most people in those countries, the apps were interchangeable. But for Musers, it was code red.

"I was scared. I was really scared," said Ariel Martin, one of the top three stars on Musical.ly. "I'd developed my whole career on Musical.ly. I was doing Musical.lys every day for years straight."

Prominent creators on the platform were warned in advance of the changes afoot. They knew about Musical.ly's big-ticket acquisition and the new app's ambitions. Company staffers explained to concerned top Musers that the app would be moving beyond lip-syncing and music. It would be a generalist app, more like YouTube but built for mobile. Users would find short-form videos related to cooking, fashion, tech, lifestyle, jobs, and more. Musers were assured that all of their content and followers would port over to the new app.

TikTok's decision to rebrand Musical.ly was a serious gamble. The relaunched app—new logo, new name, fresh marketing—was part of ByteDance's plan to break into the U.S. market.

"I was like, 'What am I gonna do? What are *they* gonna do?'" Martin recalled. "They kept being like, 'No, it's OK, you can do anything on the app now. You can do makeup on there.'" She understood the company wanted to diversify, but the pivot felt almost comically broad.

Martin eventually figured out how to create plenty of content on TikTok. But she increasingly felt that the lip-sync format she'd pioneered throughout her rise was becoming stale, even to her. She began looking for an exit route. She sought to translate her online clout to a career in the traditional entertainment industry. "Musical.ly shutting down pushed me to get into acting more," she reflected. Soon, she landed a spot in the cast of *Zombies*, a Disney Channel franchise, and she was also cast in a Nickelodeon movie called *Bixler High, Private Eye*, which came out in 2019.

While Martin still posts on YouTube and TikTok occasionally, she has largely stepped away from life online. Even at the top, to her, being a creator didn't feel secure enough. As she pivoted, she reflected on the toll that years as a top internet creator took on her mental health. "I'm a naturally very sensitive person, and social media can be toxic. People can be so terrible online. For me and my own mental health, I've had to learn not to be on it 24/7," she told me. "Every day I was presenting my entire life to the world: my friendships, my family, my breakups, things that happened to my little brother at school. It was too much. It became overwhelming."

Other leading Musers were having doubts of their own as they transitioned to TikTok. Lisa and Lena, two German identical twin sisters who had the largest Musical.ly following of all, quit the app in the months after TikTok's launch. The duo said they needed to attend to their mental health. It was too full of hate now, they said, and they didn't foresee the new app hanging onto what allowed the Musers to reign supreme.

Other top Musers embraced TikTok and were able to make the transition. Loren Gray started posting lip-sync videos in 2015 when she was just thirteen. At age seventeen, she shot to number one on TikTok after Lisa and Lena left the platform in March 2019. She had dropped out of high school because of classmates bullying her over her posts, and she focused on social media full-time. As TikTok grew, Gray embraced the platform and flipped her Musical.ly following into a TikTok empire driven by dance videos, vlogs, confessional posts, comedy, and sponsored content.

Zach King, a video-editing wizard, also made the leap from Musical.ly to TikTok. A former top Viner, he found a home on every social media platform out there. Musical.ly became one of his biggest platforms, even if his content—seamlessly edited skits that looked like special-effects magic—wasn't lip-syncing. When Musical.ly became TikTok, and TikTok embraced a wider variety of content, King was perfectly positioned. Over the course of 2019, he rose the ranks to become a top account on the new platform, with several of his videos among the most viewed of all time.

Musical.ly's impact on the internet remains. There would be no TikTok stars without Musical.ly. The app ushered in an era of mobile-first video editing tools that other platforms struggle to replicate to this date. It got an entire generation, Generation Z, comfortable with sharing and posting videos of themselves publicly online.

"Musical.ly made social media more accessible to the average person," said Ariel Martin. "It's hard to remember, but back then, with YouTube you had to have a camera and editing software. Vine you just had six seconds to do something funny—it was all comedy. Musical.ly was a completely different way of expressing yourself. You could kind of become whoever you wanted."

Despite consternation among Musers, TikTok's huge marketing efforts were paying off. Throughout its first year, TikTok steadily added new users around the world. A new generation of stars emerged on the app. Charli D'Amilio, currently one of the largest TikTokkers in the world, signed up for the app in May 2019. Her rise was so swift she became a meme. Addison Easterling, now among the top five most followed accounts, joined in July 2019. Charli's older sister, Dixie D'Amelio, eighteen years old, was also on the rise. Their content was vintage Musical.ly—lots of dances and lip-syncing. They rocketed up the charts, gathering tens of millions of followers in the next year, driven by the power of TikTok's "For You" page.

The "For You" page was TikTok's biggest innovation as a social media platform. The page uses AI to create an algorithmically programmed feed that serves users content based on what they are likely to find most engaging. Every other social media app prioritized posts from the accounts the user followed. TikTok's recommendation engine didn't care who you followed. It would learn simply by noting what you stayed to watch and what you swiped past. If you lingered on a particular sub-genre of video, the app would serve you more of it. For creators, if you posted something engaging enough for people to watch, the algorithm would promote it, distributing your content to millions within minutes, even if you didn't have a single follower.

This innovation sped up the cycle of virality to a breakneck pace, the mechanics of which were bewildering to some observers. "With every new TikTok star who dances or smirks their way to a million followers come just as many more people asking why they deserve to be famous in the first place," Rebecca Jennings wrote in *Vox*. "The cycle of overnight fame and equally swift backlash is going to keep happening."

One of the best examples of the TikTok algorithm in action was the case of Bella Poarch. In April 2020, the twenty-three-year-old Philippine American began experimenting with TikTok videos, mostly posting cosplay clips and lip dubs. By the beginning of August, she'd gathered a healthy 100,000 followers on the app. Then, on August 17, she posted a short clip lip-syncing to Millie B's "M to the B (Soph Aspin Send)" and making expressive faces for the camera. It was just ten seconds long, and hardly seemed to most viewers like a standout video. But it took off like wildfire, boosted worldwide by the "For You" algorithm. Within a day, Poarch gained 2 million followers, and the clip became the most-liked video on TikTok ever. She kept posting lip-sync content and continued to gather nearly 1 million followers per day, reaching over 22 million followers just one month after the original viral post, launching her into the top twenty accounts on the app.

Inevitably, Poarch's rise begot backlash from cringe and drama channels. A string of minor controversies kept her in the conversation until Poarch had accrued both hardened critics and fans who felt that she was being wrongfully attacked. It was the perfect recipe for fandom, and from there it was almost as if her celebrity had reached critical mass to support itself. Years later, Poarch's star has only continued to rise. At present, she is a top-three account on TikTok, and she's converted her rapid rise into multi-platform success, launching successful YouTube and social accounts and releasing hit singles.

※

By the end of 2019, the old Musical.ly guard were no longer the hottest stars on the app. The D'Amelio sisters, Easterling, and several dozen other Gen-Z stars had become internet A listers and that December, on Christmas Day, they made an announcement that defined the early TikTok era. Dressed in matching jeans and white T-shirts, fourteen creators huddled together against a white backdrop making goofy faces. The stars had come together to launch the Hype House, a new creator collective and content house.

Collab houses (also known as content houses) were by now a long-running tradition in the influencer world. The first content house had been the Station, which YouTubers operated in Venice Beach. In 2014, members of a YouTube collab channel called Our Second Life moved in together in Los Angeles in what was known as the O2L Mansion. In 2015, nearly all the top talent on Vine moved into the 1600 Vine apartment complex. YouTuber mansions dotted L.A. by 2017, with members of the Vlog Squad living in Studio City, Clout Gang renting a $12 million mansion in the Hollywood Hills, and Team 10, Jake Paul's infamous YouTuber collective, occupying a giant house in West Hollywood before eventually decamping to a mansion in Calabasas.

The Hype House's Backstreet Boys–esque photo shoot exploded online, trending within hours of being posted. The group's handle racked up millions of followers in under a week. It was all any TikTok users under the age of eighteen seemed to be talking about. Videos including the hashtag #hypehouse accrued nearly 60 million views on TikTok in its first few days of existence. Just over a year later, that number would reach 7 billion.

The masterminds of the Hype House were Chase Hudson, age

seventeen, a TikTok star known online as "Lilhuddy," and Thomas Petrou, age twenty-one, a YouTube star and former member of Jake Paul's Team 10. The duo had a combined 40 million followers. In November 2019, they began plotting a move along with other members of the house and signed a lease for an Encino mansion just two weeks later.

The physical Hype House was a Spanish-style mansion perched at the top of a hill on a gated street in Encino, with a palatial backyard, pool, and massive kitchen, dining room, and living quarters. While the Hype House originally consisted of seventeen members, only a few members actually lived there full-time: Chase Hudson, Thomas Petrou, Daisy Keech, Alex Warren, and Kouvr Annon. Others, including Charli and Dixie, kept rooms to crash in while they were in town. And, day by day, a steady stream of influential internet stars, including James Charles and David Dobrik, stopped by to pay homage to the new guard.

The Hype House was a boon to every member of the group, but if you wanted to be part of it, you needed to churn out content daily. "If someone slips up constantly, they'll not be a part of this team anymore," Thomas told me in 2019. "You can't come and stay with us for a week and not make any videos. It's not going to work." This steady stream of content fueled TikTok, and created a feedback loop to Hype House members of more followers and views. It helped cement TikTok as *the* app of its generation.

"It's a brilliant move for power players on these platforms to lift each other up," explained Sam Sheffer, a YouTuber and technologist. "'Elevate others to elevate yourself' is a saying, and it really rings true with this new generation of TikTokkers."

As Hype House took off, dozens of collab houses began cropping up across Los Angeles, often to the chagrin of locals. While some collab houses were collectively owned, others were run by management companies, a new model for creators that took a page out of the Silicon Valley incubator playbook: Houses run by management companies would take a percentage of the revenue earned by each creator living there. In these houses, social media stars would pay discounted rent with money that they earned from brand deals. The management companies were supposed to cover everything else, including the remainder of the rent, utilities, cleaning staff, as well as marketing, PR, and legal services for every member.

In January 2020, less than a month after Hype House launched, a group of young men aged eighteen to twenty known as the "Sway boys"—which were often called "the One Direction of TikTok"—moved into a 7,800-square-foot mansion on a quiet street in Bel Air. The house was dubbed the Sway House. The group comprised some of the biggest male Gen-Z TikTok creators on the internet: Bryce Hall, Jaden Hossler, Josh Richards, Quinton Griggs, Anthony Reeves, Kio Cyr, and Griffin Johnson, all of whom were represented by TalentX, a management company. TalentX leased the house, and the residents lived there for free in exchange for TalentX getting a cut of their brand deals. The Sway boys were loud, often playing music late at night and hosting wild parties. The neighbors eventually drove the group out of the area, and several of the Sway boys rented a new home in the Hollywood Hills.

Such controversy did little to slow the trend, though. New collab houses popped up all over L.A., then America, and then the world. The House That Nobody Asked For, a content house in Las Vegas, became popular among alternative kids. Gen-Z houses in Los Angeles like the Vibe House, the Alpha House, the Drip Crib, and more arose. After the Hype House was criticized for not having any Black members, some TikTokkers discussed forming a Melanin Mansion.

Pretty soon the original Hype House fractured and multiplied. FaZe Clan, a collective of internet-savvy gamers, took over Justin Bieber's former home in Burbank and turned it into a gamer's mecca. The Hype House moved into the old FaZe Clan house in the Hollywood Hills, while Daisy Keech, a Hype House co-founder, splintered off and co-founded a new house called Clubhouse.

Clubhouse took the incubator model a step further, becoming the first publicly traded content house in November 2020. This suddenly made it possible for fans to buy a stake in a top content house. Within two years, however, the stock plummeted, becoming practically worthless.

Brands also got in on the content-house business. The ecommerce platform Wish and the social media app Clash both rented sprawling mansions where creators could produce exclusive content. Other brands did pop-up collab houses around L.A.

As the competition among young influencers in Los Angeles intensified, and as TikTok collab houses sucked up attention, YouTubers who had previously felt secure in their status as internet elites were suddenly threatened by the new wave of talent.

"TikTok has brought a younger group of creators. That energy is kind of pushing on a lot of older creators," said Josh Sadowski, age nineteen, a TikTokker with nearly 4 million followers who lived in a collab house. YouTube stars began frantically reaching out to Gen-Z TikTokkers, hoping to team up on videos. They could feel a changing of the guard. Suddenly, huge YouTubers were begging to collaborate with up-and-coming TikTokkers. In October 2020, Dave Portnoy, the forty-five-year-old founder of Barstool Sports, seeking relevance with a younger audience, inked a partnership with Sway boy Josh Richards for a podcast where they discussed Gen-Z TikTok drama.

Traditional Hollywood also sought to capitalize on the collab house boom. Houses full of beautiful young people were a reality TV staple, and these particular young people came with the bonus of an enormous, preexisting fan base. One event in particular, during the first week of July 2020, captivated execs at several big Hollywood production houses. Naturally, it involved drama between two of the biggest collab houses.

Late one Monday night, members of the Sway House barreled up the streets of the Hollywood Hills toward the Hype House with the intention of confronting fellow TikTok star Chase Hudson, after he'd made comments about the Sway boys online.

The late-night raid was the culmination of a weekend full of non-stop drama between some of TikTok's biggest stars, in what became known as the "TikTokalypse." Tensions between friends that had been simmering for months boiled over in the form of TikTok videos, quote tweets, Instagram Live rants, and lengthy Notes app commentary, a multimedia avalanche of accusations of cheating and betrayal (business and personal). Top YouTubers like Tana Mongeau immediately recognized an opportunity to insert themselves into the conversation and fanned the flames by reacting in real time.

As soon as the Sway boys stepped out of their car that Monday night, they were swarmed by paparazzi. The Hype House members quickly ushered them into their 14,000-square-foot home and squashed the beef in private. "we talked. no fighting. it's settled," Jaden Hossler, nineteen, tweeted.

"This is a TV series waiting to be made," a film producer named Hemanth Kumar tweeted about the TikTokalypse. "Who's calling dibs on this one?"

Suddenly, every major TikTok collective was on the reality show hunt. Wheelhouse, a production studio, partnered with Hype House to produce a reality show called *Hype House* for Netflix. The D'Amelio family landed a reality show on Hulu called *The D'Amelio Show*. Clubhouse partnered with talent agency ICM to shop around its idea of a show, though it didn't take.

What traditional Hollywood (which almost uniformly continued to think of TikTok as a teen dance app) still didn't realize is that social media is itself a 24/7 reality show, delivered to young people in a format they're far more interested in consuming than traditional TV. With TikTok, the reality show culture of the aughts and traditional fame had finally merged. Young viewers weren't interested in ham-handed storylines about their favorite influencers delivered on a streaming service six months later. They could watch it all play out in real time on the internet, for free.

As the TikTokalypse happened, millions of young fans turned to @TikTokRoom, an Instagram account founded by two teenagers named Elasia and Nat in 2015. The page was modeled after The Shade Room, which chronicled celebrity news on Instagram. Elasia and Nat's original page shared news about Musical.ly stars, but when TikTok took over, they revamped it (and renamed it). Posting dozens of times a day, TikTokRoom regularly broke news. Cece Price, who ran the Instagram drama news page @M3ssyM0nday, also gained traction during TikTokalypse. For a while the two pages were neck and neck in followers, both surpassing over half a million.

Gen-Z has its own reality TV and app, live from Hollywood, no television screen or entertainment execs needed.

// CHAPTER 20 //

Unlocked

In March 2020, the Covid-19 pandemic shoved everyone indoors. The world was becoming ever more online, but now the holdouts' hands were forced as the internet became the main portal to other people. We experienced political and cultural upheaval staring at our phones. Views for online creators of every stripe skyrocketed. A huge spike in downloads of apps like TikTok and Twitch dramatically broadened their user bases. As TV hosts broadcast from home, there was now production value parity between traditional entertainment and social media. Late adopters finally recognized online creators as legitimate.

By this point, social media was already ingrained in daily life for many. Service workers, for example, had already embraced TikTok. Stuck on long shifts with much downtime, many such workers turned to the app to give others a glimpse of their lives. When the service industry was hit hard by the pandemic, these workers leveraged TikTok to promote fundraisers for themselves and their employers.

TikTok also played a key public health role in the early days of the pandemic, serving as a platform for frontline workers to educate the public on Covid based on firsthand experience. ER nurses, doctors, and health workers used TikTok to explain why masks were important, why Covid-19 was so dangerous, and to debunk vaccine misinformation.

Sick people chronicled their health journeys and used the platform to connect with audiences of millions online from their beds.

The increased popularity of cooking and baking during the pandemic was also driven by social media, especially TikTok. The platform popularized viral food trends like pancake cereal and whipped coffee. It also birthed a new generation of food creators like the eighteen-year-old darling of the food world, Eitan Bernath. When schools ceased in-person education, students and teachers turned not just to Zoom video conferencing but also to TikTok and YouTube to aid in remote learning. Students hosted live-streamed study sessions where they helped each other with homework.

Stuck inside, young users staged imagined lives on social media. Roleplay was not new online, but the performances during the pandemic were particularly creative. Young people constructed elaborate fake story lines and scenarios as part of POV (point-of-view) videos. Thousands of teenagers began cosplaying multinational corporations and forming absurdist cults centered around their favorite TikTok creators. Book recommendation videos under the hashtag #BookTok, created mostly by women in their teens and 20s, made blockbuster bestsellers out of books new and old.

During a period of fear and isolation, online connections became more essential than ever. The vibrancy and effervescence of the internet in 2020 gave millions of people a sense, if only for a few minutes at a time, of normality and optimism.

※

OnlyFans took social media from the Paul Smith wall to the bedroom. If Vine, Instagram, and TikTok had succeeded at making users feel like they were getting a glimpse into a celebrity's private life, OnlyFans tempted with a glimpse of something much more intimate. Combining the porn and sex work industry with the rapidly expanding online creator world, OnlyFans took off by taking advantage of the powers that made social media so revolutionary: the ability to bypass traditional gatekeepers and to monetize one's own content directly. OnlyFans empowered adult performers, who had previously been at the mercy of adult entertainment companies, to create and monetize their own images.

OnlyFans was founded in London in 2016 by brothers Tim and Thomas Stokely. From the start, the focus was on pornography by subscription. Performers would keep 80 percent of their subscription earnings and 100 percent of what they made from merchandise sales.

Unlike other social media platforms, OnlyFans does not have an app, due to Apple's and Google's bans on pornographic content. Still, the site has flourished, and grew especially fast during the pandemic. OnlyFans' pre-tax profits increased from $61 million in 2020 to $433 million in 2021. Total active users increased 128 percent in the same period to almost 188 million, while the number of performers increased by a third to just over 2 million. Those performers earned a collective $4 billion, more than double the prior year.

Many of those performers attempted to promote themselves on other social platforms, especially TikTok, Instagram, and YouTube. They also cultivated relationships with online porn and fetish communities. But it was tough work to build a follower base from scratch, and a thousand times more difficult when you were a sex worker. Mainstream social media platforms, who have made it clear that adult content is not welcome, frequently delete or shadow-ban sex workers, denying them valuable promotional channels.

It would have been naïve to think that the vast majority of sex workers on OnlyFans would be able to financially prosper, despite the amount of money the site was making. In 2021, *Fast Company* reported that OnlyFans content creators had earned $3 billion since the launch of the site, but there was no clear breakdown of how much the average performer brought in. That same year, an OnlyFans spokeswoman told Charlotte Shane of the *New York Times* that "over 300" creators had been paid out "over $1 million."

For some creators, however, OnlyFans brought about an unprecedented liberation. OnlyFans was part of a broader shift, starting in early

2020, of creators making money directly from fans through subscription revenue. The economic upheaval of the pandemic made many realize the volatility and unreliability of advertising revenue, and creators began diversifying revenue streams away from sponsored content. At the same time, the rise in perceived value of online creators encouraged more users to pay for digital content.

Patreon, which allows creators to charge monthly subscription fees to fans in order to access exclusive content, experienced a massive upswing. Over 30,000 creators joined Patreon in March 2020 alone, *TechCrunch* reported. By September the company had raised a fresh round of funding valuing the company at over $1.2 billion.

As it would turn out, the early pandemic Patreon and OnlyFans bumps were temporary. But they set the stage for a wave of new monetization schemes. By 2022, dozens of venture-backed "creator economy" start-ups were making it easier than ever for anyone to monetize any aspect of their life—down to what they ate or who they hung out with. Tools aimed at facilitating small payments and revenue sharing made it easier to micro-monetize a person's every online move.

※

Teenagers with large TikTok audiences weren't just posting dance and comedy content in 2020, they were using the platform for political commentary and activism. This was the year of Covid, but also the pivotal showdown between presidential candidates Donald Trump and Joe Biden; the year George Floyd was murdered at the hands of the Minneapolis police.

Content on TikTok had been getting more political for a while, and it was clear that the 2020 election would be the first "TikTok election," with the majority of young people turning to the app for news and real-time information on issues. Throughout 2020, content creators attracted millions of views by discussing abortion, gun control, climate change, and universal healthcare. Creators who produced news and chatting channels saw their viewership explode. "Just Chatting," a category on Twitch where creators discuss various topics with their audience, ballooned, launching the careers of Hasan Piker, for example, a left-leaning influencer who covered news and politics.

Teenagers also began to form online political collab groups, which they dubbed "hype houses" after the famous collective. On the right, such groups included @conservativehypehouse, @theconservativehype house, and @TikTokrepublicans. A group called the @therepublican hypehouse amassed more than 217,000 followers in less than a month, and then blew up to over 1 million followers before being banned for spreading election-fraud conspiracies. On the left, notable houses included @liberalhypehouse and @leftist.hype.house. In the middle, there was a bipartisan hype house and a smaller account called @theneutral house.

Everyone got political and found a community online for their interest or niche. Online hype houses for religions like the Jewish hype house and the Muslim hype house emerged to speak to specific religious communities. There were hype houses for every state (Michigan alone had at least three), many major cities, and various colleges. Cabin Six, an LGBT-focused TikTokker collective, held public auditions on TikTok in early 2020, as did Diversity University, another TikTok collab group that organized a pop-up house in L.A.

These groups spoke to specific voter cohorts, and they filled a news vacuum that the mainstream media was once again ignoring: young people who wanted information tailored to their specific background and experiences. TikTok influencers were only too happy to swoop in. Their work had real influence on the broader political conversation. LGBTQIA+ TikTokkers raised awareness about Mike Pence's track record curtailing gay rights by promoting a meme suggesting that he would send them all to gay conversion camps. The nickname "Mayo Pete" for Pete Buttigieg, which tainted his public image at a crucial moment for his campaign, was born out of TikTok. And geopolitical events like the threat of nuclear war with Iran and President Trump's first impeachment became talked-about news stories among Gen Z after TikTokkers launched widespread memes on the app.

Social media platforms served as a hub for on-the-ground coverage of Black Lives Matter (BLM) protests throughout the country. After seventeen-year-old Darnella Frazier filmed the murder of George Floyd and uploaded it to Facebook and Instagram, setting off a wave of protests, Twitch streamers and TikTok creators began covering BLM protests. One TikTok of a BLM protest in Minnesota, shot by creator Kareem Rahma, got over 54 million views, a level of reach and virality no traditional cable news channel could rival.

In late June 2020, TikTok users banded together to register thousands of tickets for a Trump campaign rally with no intention of attending, leading Trump to deliver a speech in a large, sparsely filled arena. The stunt garnered mainstream attention and proved that TikTok could be used for mass mobilization around political causes.

Not coincidentally, 2020 was also the year TikTok was forced to contend with rampant disinformation. In October of that year, the platform finally cracked down on QAnon and Pizzagate conspiracy theories. However, other conspiracies, such as one falsely alleging that furniture seller Wayfair was engaged in child trafficking, and other theories related to sex trafficking, seized the app for days.

As the election loomed and Trump threatened to ban TikTok, a group of TikTokkers banded together to form "Gen Z for Biden," a collective of hundreds of top TikTok stars aimed at encouraging young people to vote Trump out of office. When states were called on Election Night, millions of teenagers followed the results on TikTok. After the election, "Gen Z for Biden" became "Gen Z for Change," and began working closely with the Biden administration to push key parts of the president's agenda, such as the vaccine rollout and child tax credit. By the spring of 2022, the administration was holding press briefings for TikTok creators on major news events like the war in Ukraine.

※

Thanks to social media, millions of marginalized people went online and found community, purpose, and even new income. But if the internet was reshaping the world, it was also of the world. Hateful people, including far-right extremists, could follow the influencer formula too. "Content is king for these people," Ciaran O'Connor, an analyst at the Institute for Strategic Dialogue, explained in the Southern Poverty Law Center's "The Year in Hate and Extremism Report. "That's what drives this: they just want to create content that they can share online, they can get donations and views off of, [and] they can share later on as clips."

More insidious bias could be found within the creator community, too. It usually involved using someone else's innovation to boost your own profile—and potential earnings. As was the case in the offline world, such appropriation often came at the expense of Black creators.

In 2020, a dance nicknamed "the Renegade" exploded across the internet. It was a massive viral hit, and uncountable teenagers could be seen doing the Renegade in public and online, where they furthered the spread. But it wasn't just teenagers: stars like Lizzo, Kourtney Kardashian, and members of the K-pop band Stray Kids all performed it.

The one person who was not able to capitalize on the attention as the dance ascended was Jalaiah Harmon, the fourteen-year-old girl in Atlanta who created it.

On Sept. 25, 2019, Harmon came home from school and asked a friend she had met through Instagram if she wanted to create a post together. She listened to the beats in the song "Lottery" by the Atlanta rapper K-Camp and then choreographed a difficult sequence to its chorus, incorporating viral moves like the wave and the whoa. Harmon filmed herself dancing the choreography and posted it, first to a smaller app similar to Dubsmash, a short-form video app, and then to her more than 20,000 followers on Instagram (with a side-by-side shot of her and her friend).

Within weeks, the dance began to take off among users of Dubsmash and Instagram, and shortly after, on TikTok too. The biggest creators on TikTok began posting the dance, but none gave Harmon credit. After long days in the ninth grade and between dance classes, Harmon tried to get the word out about her authorship of the dance. She hopped in the comments of several videos, asking influencers to tag her. She was universally ignored.

Cross-platform sharing—of dances, of memes, of information—was common for creators trying to grow their following. But to be robbed of credit is to be robbed of real opportunities. Creators of popular dances, like Backpack Kid or Shiggy, could amass large online followings and become influencers themselves. That, in turn, opened the door to brand deals, media opportunities, and career options.

For Dubsmashers and those in the Instagram dance community, it was common courtesy to tag the handles of dance creators and musicians, and use those hashtags to track the evolution of a dance. Within TikTok, not so much.

On January 17, tensions boiled over after Barrie Segal, the head of creators at Dubsmash, posted a series of videos asking TikTokkers to give a dance credit to D1 Nayah, a popular Dubsmash dancer with more than 1 million followers on Instagram, for her Donut Shop dance. @TikTokRoom picked up the controversy and spurred a sea of comments.

After I reported on the dance's origins in a story for the *New York Times*, Jalaiah was finally given the credit she deserved. She was showered with brand deals and opportunities. Her experience set off a wave of discussion around credit and appropriation in the creator industry. Despite creating and driving many of the internet's biggest trends, Black creators received fewer brand deals and were consistently paid less than their white peers. Accounts like @InfluencerPayGap called attention to the pay disparities between Black and white influencers. Black creators were featured less frequently in brand campaigns and received inferior treatment at brand-sponsored events. Black founders have also been disproportionately overlooked when it comes to venture capital.

Not all that far from where Jalaiah Harmon created the Renegade, Keith Dorsey, a talent manager based in Atlanta, watched the events of 2020 unfold with great interest. He saw the boom in content houses but realized that there had yet to be a mainstream all-Black creator house, and so he sought to build one.

By summer 2020, many of the young TikTok stars Dorsey managed had graduated high school and were anxious to escape their parents' houses, desperate for the in-person socializing that they'd been robbed of in quarantine. Modeled after the collab houses that had opened in L.A., Dorsey and a group of creators he managed launched the Collab Crib, an all-Black content house in Atlanta. Around that same time, two dozen Gen-Z creators formed the Valid Crib, another all-Black content house across town.

Harmon's Renegade experience revealed a small piece of a much larger and ominous cultural picture. Collectively, the members of the Atlanta collab houses were responsible for dozens of viral trends. They were also regularly featured on massive meme and Instagram accounts like Worldstar and the Shade Room. But for both groups, sponsorship was a challenge. Several brands dangled opportunities that failed to materialize, with one potential backer instead choosing to put money behind an all-white creator house. A home furnishing company responded that the demographic wasn't a fit for their brand and refused to even take a call.

As protests for racial justice swept America, people also began scrutinizing the influencer industry, which had long been dominated by attractive, straight, white, young people.

"Racism even plays into the algorithm and why Black creators tend to have smaller followings," said Chrissy Rutherford, a digital creator and a founder of 2BG Consulting, a brand consultancy focused on diversity and inclusion.

"We are being exploited, and that's the core issue Black folks have always had in terms of labor," Erick Louis, a Black content creator, told me. "These millions of likes, that should all translate to something. How do we get the real money, power and proper compensation we deserve?"

Black creators began speaking out and demanding credit for their creative work, whether it was dances, catchphrases, or viral memes. Suddenly, brands began casting Black creators in ad campaigns and inviting them on influencer trips. But it didn't last. By mid-2023, the industry had almost fully rolled back the progress it made in terms of diversity.

The Collab Crib eventually switched to a studio model, where members could meet up to work before returning home. The Valid Crib shuttered just a year after launching. Many Black creators were still confident that they would succeed together, but everyone knew the hill was steeper than it should have been.

"A lot of brands have complained that there's no Black creators with larger followings," Rutherford told me at the time, "but it's like, have you ever considered that you are basically only engaging and following and giving likes and opportunities to white creators? It works both ways."

// **CHAPTER 21** //

The Scramble and the Sprawl

MANY INFLUENTIAL SILICON VALLEY INVESTORS HAD, FOR YEARS, dismissed the online creator world, writing off influencers as silly and frivolous. But when the pandemic hit, the new landscape became obvious. Feeling the ground shift, these investors began funding seemingly any start-up catering to online creators. The "creator economy" became a buzzword and entered the Silicon Valley hype cycle. Some investors sought to manufacture their own online influence by leveraging the apps they funded.

The frenzy around the creator industry among Silicon Valley VCs, along with TikTok's sudden success, revived the consumer social sector, which had previously been stuck at a standstill. "Covid happened, and that was a huge inflection point," said Li Jin, a venture capitalist. "It felt like everything was up for grabs again. . . . and the rise of TikTok made everyone realize, oh, there can be innovation in consumer social. What TikTok did really differently was cater to creators as first-class citizens."

But the ignorance of some in Silicon Valley was pervasive.

Because many men in tech associated the word "influencer" with women and wanted to distance themselves from past statements dismissing influencers, they pushed the terms "creator" and "creator economy" to reflect the now massive, dominant industry. Their ignorance was such that they sought to fund companies they believed were first movers in a market, only to learn they were competing with established funders in the space. For instance, one Silicon Valley exec expressed interest in investing in referral-code platforms with e-commerce capabilities, seemingly unaware of RewardStyle.

Tech news site the *Information* estimated that in the first six months of 2021 alone, venture capital firms invested over $2 billion in at least fifty creator-focused start-ups. As the year marched on, those numbers expanded. Perhaps the most egregious valuation to come out of this era was Clubhouse (unrelated to Daisy Keech's content house, which had flamed out quickly), a live audio app that allows users to host real-time chat conversations or listen in on others' real-time chat shows. Some Silicon Valley executives, desperate to replicate the TikTok boom with a new social platform, plowed millions into Clubhouse. In April 2021, Clubhouse raised $200 million in a funding round led by Andreessen Horowitz, putting its valuation at roughly $4 billion. Major platforms like Spotify, Facebook, and Twitter rushed to copy Clubhouse.

Clubhouse itself, however, flopped hard. The app's investors used the app to promote their own personal brands, attempting to turn themselves into influencers and alienating the app's wider user base in the process. Clubhouse also angered popular creators by refusing to institute basic user safety measures. Female creators on Clubhouse dealt with vile abuse, stalking, and doxxing as the platform's leaders ignored and dismissed their concerns. The app would be irrelevant within a year.

Recognizing how out of the loop many in Silicon Valley were with online culture, some influencers set out to become venture capitalists themselves. In March 2021, Jake Paul announced his second attempt at launching a venture capital firm. A month later, TikTok stars and former Sway boys Josh Richards, Griffin Johnson, and Noah Beck teamed up to start their own VC firm called Animal Capital.

Other creators sought to ride the funding wave with start-ups of their own. In February 2021, David Dobrik raised $20 million for Dispo, a camera app and social platform that mimicked a disposable camera. The company was valued at $200 million, and the app became one of the year's most-hyped new social platforms. Just a month later, however, an investigation by the journalist Kat Tenbarge for *Insider* revealed sexual-assault allegations regarding a member of Dobrik's Vlog Squad, his YouTube prank ensemble. Tenbarge's revelation set off a wave of backlash against Dobrik. Dispo's investors cut ties with him and said they would donate all profits from the investment to an organization working with assault survivors. Dobrik issued multiple public apologies to his fans. This was yet another reminder that the large audiences that made some creators famous could later bite back.

※

By 2022, the creator industry was mainstream and massive enough that even small and mid-level talent knew not to work for free. Platforms and tech executives soon realized that if they wanted a piece of the online creator pie, they'd need to pay influencers, so that's what they did, sometimes exorbitantly. Before it stopped paying its bills and stiffing dozens of Black creators, Triller, a short-form video app, doled out money to TikTok's top stars. Triller leased a white Rolls-Royce with a "TRILLER" vanity plate for Charli D'Amelio, as well as a Mercedes-Benz for Sway boy Josh Richards.

Snapchat also began showering money on creators, launching a TikTok clone called Spotlight and paying out $1 million a day to the creators who got the most views. Twitter allowed creators to put their content behind a paywall and charge monthly subscription fees. Facebook, once again, sought to court creators, offering millions of dollars to top influencers to use products such as Instagram Reels, the company's TikTok clone.

Soon tech giants adopted a more traditional way to pay talent that was similar to the system that YouTube pioneered over a decade earlier: ad revenue sharing. Instagram announced a program to split advertising revenue with Reels creators. TikTok announced a program called Pulse, where it would run ads against the top 4 percent of content on the platform and share ad revenue with the creators. In early 2023, YouTube introduced revenue sharing on Shorts, its TikTok clone. Social media giants also added features that let users "tip" creators, building on a trend of direct monetization led by platforms like Patreon, OnlyFans, and the email newsletter publishing platform Substack.

With more people than ever shopping online during the pandemic, social media was minting new influencers faster than brands could keep up. In response, many creators turned to affiliate marketing to monetize. Relatively early in its existence, Amazon had welcomed third parties to help drive sales; then, in 2017, it launched its "Amazon influencer" program, aimed at helping content creators launch their own Amazon storefronts, from which they'd receive a small amount of money for every product sold. Now, Amazon was speedily onboarding thousands of new TikTok content creators, each given a unique vanity URL on Amazon to make it easy for followers to remember and find their storefronts.

Walmart launched its own influencer program in 2022, announcing, "Anyone can be a creator." As with Amazon, creators in the Walmart program were given access to tens of thousands of products, which they could earn commission on. These shifts, along with the rise of e-commerce platforms and drop-shipping, have made it easier than ever for influencers to become ecommerce power houses.

In 2023, despite a slowdown in funding of "creator economy" start-ups, the economic power of creators only continues to grow. Goldman Sachs predicts that the market size of the creator industry will double to half a trillion dollars by 2027. As a result of this shift, individual creators now aspire to financial results that were previously exclusive to large corporations. Influencers Emma Chamberlain and Charli D'Amelio have launched their own coffee and footwear lines, respectively. The Kardashians have leveraged their social media reach to produce several billion-dollar companies, including Skims and Kylie Cosmetics. In 2022, YouTuber Jimmy Donaldson, who goes by MrBeast, was projected to earn $110 million across his channels. "A lot of people still see YouTubers as a subclass of influencers," Donaldson told Forbes. "They still just don't truly understand the influence a lot of creators have." Online creators have flipped the business script. Whereas traditional companies would develop a product and then find a way to market it to consumers, creators build an audience first and then develop products tailored to the fans they already have.

The explosive growth of the creator industry is the culmination of decades of user-driven platform evolution. The internet connected the world, allowing talented creators to bypass traditional gatekeepers and build fanatical audiences directly. When one thinks of "the media," they often think of broadcast news and newspapers; in reality, creators are "the media" of today. The media landscape that they dominate is only becoming more digital and more distributed. Cycles of virality are accelerating. Online influence can make you an overnight Hollywood sensation, morph you into a powerful business leader, or take you to the White House. These shifts will only be compounded with technological advances such as the rise of AI. Legacy institutions that refuse to adapt will continue to fade into oblivion.

// EPILOGUE //

WE'VE SEEN HOW MUCH THE INTERNET HAS CHANGED IN THE TWENTY-first century so far, and how much we've changed as a result. This transition is accelerating as the online and offline worlds merge. A lone user can redirect a platform overnight, and technology founders almost never foresee how their creations end up being used. Over the past two decades, an unprecedentedly innovative community has managed to refresh itself with every iteration of the social media landscape. Online creators don't just produce content; they define the norms and dynamics of their medium.

As we look to the next generation of technology products, users must recognize the power they hold and not surrender decision-making to Silicon Valley executives. We have the chance to create a better system that amplifies independent voices and jettisons the flaws of traditional media and legacy institutions. Tech founders may control the source code, but users shape the product.

Our story began when the internet lowered the barrier to publishing, allowing independent authors to gain a following directly, and serve communities who were previously overlooked. Social platforms emerged and lured average people online, teaching them to post for an audience. As the platforms scaled, they introduced public metrics, rolled out new content formats, and attracted advertisers, laying the groundwork for users to redefine fame and take advantage of significant new economic opportunities. Platforms that partnered generously with their top users—YouTube first and most notably—have been rewarded handsomely. Platforms that neglected—or worse, fought—creators, did so at their own peril. As creators competed for eyeballs, the attention arms race led to the extremification of content, volatility of revenue streams, and personal burnout. That momentum only intensified as TikTok massively expanded in the United States, the pandemic brought even the holdouts online, and the "creator economy" forced the business world to rewrite its playbooks.

The rise of social media and the creativity of its users has given more people the chance to benefit directly from their labor than at any other time in history. This expansion of opportunity has been particularly life-changing for many who have been historically shut out of legacy institutions. The creativity and tenacity of online creators has challenged traditional gatekeepers as never before, often with socially and economically liberating results.

At the same time, a quarter-trillion-dollar industry has emerged out of nowhere, with almost no guardrails or protections for the workers within it. We are all pressured to commodify ourselves, our lives, and our relationships in increasingly invasive ways. And we face a grave contagion in the form of disinformation and hate by influencers who otherwise would have been constrained by the limits of budget and public access.

Content creators are the new media. No matter how hard you try to avoid it, you're in their online world too. Those without active social media accounts have information about them posted online by others, creating a digital reputation more widely accessible than the offline version. This information comes from digitized public records, from friends and family who post photos and videos, newspaper articles, and school yearbook photos. This flood of online content has been enabled by technological developments that have collapsed barriers to high-quality content production. We've gone from clunky 35mm film cameras to multi-lens, AI-powered smartphones that can store effectively infinite images and videos, all of which can be instantly viewed, edited, shared, attached, and posted. The internet has made all the world a stage more than ever before.

This has led to bizarre changes in how we live: the pursuit of shareable content often seems more urgent than the desire to actually do the thing that will be recorded and shared. To take a particularly extreme example, many of the January 6 insurrectionists seemed more interested in documenting their violent ransacking of the Capitol than they did in overthrowing American democracy—their desire for online celebrity providing law enforcement ample evidence for subsequent prosecutions. That desire to document and broadcast makes sense if we understand that, for more and more people, the online world is often more "real" than our material one.

There is an underlying human desire behind this transformation. We want our existence validated, and increasingly, an online presence is the measure of that validation. And how does someone validate themselves online? By connecting, and by being noticed for the content that they're producing. That act of connection is inherently optimistic. It's a vote for the belief that being connected is better than being alone. Sometimes our connections can be intimate and direct—yourself and a friend. They can also be bonds built on respect or shared affinity. The most powerful online creators build such a bond between themselves and those watching their videos, reading their posts, listening to them talk. They advise, perform, misbehave, sell, display and they do it in a way that makes a stranger think, *This is someone I want to connect with*. And maybe, fingers crossed, that stranger will want to connect with you.

In these moments, we realize how inextricably interconnected we've become thanks to the internet. While Big Tech has consolidated power, often against the individual, preying on your privacy, content, and attention, we should heed the lessons of the first twenty years of online life, and reflect those learnings in our work to build a better internet. In this we must all be creators, influencing the online world we inhabit.

// ACKNOWLEDGMENTS //

This book would not be possible without all of the brilliant and creative people who shaped the early internet. Thank you to every source who shared their memories with me and helped to tell this story. The online world is sprawling, beautiful, and endless. There were so many stories that got cut for space that I hope to still be able to tell someday.

A deep and heartfelt thank you to Stephanie Frerich at Simon & Schuster who worked tirelessly to bring this book to life. I am so grateful for her endless patience and willingness to be available in every time zone. Pilar Queen, my agent, was a fierce advocate for this project and provided endless crucial guidance to me as a first-time author. Huge thank-you to Jon Cox and Geoff Shandler for their thoughtful editing, Andy Young for exhaustive fact checking, and Franck Germain for working nights and weekends pulling together citations.

Thank you to everyone in the tech, entertainment, and media industries who spoke to me on background and read early versions of this book. A special thank you to Brendan Gahan for his bottomless generosity and going above and beyond to help me in this process, Tim Shey for being such a helpful resource, and Joshua Cohen for being so kind and helpful in connecting me with sources over the years.

Technology reporting is a heavily male-dominated field, but it's women reporters covering the online world and creator industry whose work most informs my own. Katie Notopoulos's writing and creativity in telling stories set the blueprint for this beat. I could not have put together this book without Kat Tenbarge's internet-altering scoops, Rebecca Jennings's thoughtful analysis pieces, EJ Dickson's extensive reporting on online sex work and OnlyFans, Kaya Yuriff's daily reporting on the creator economy, Kalhan Rosenblatt's expert trend explainers, Kate Lindsey's must-read newsletter, Morgan Sung's influencer coverage, Kelsey Weekman's unforgettable online culture pieces, Daysia Tolentino's fantastic YouTuber coverage, Amanda Siberling's of the moment tech news, Madison Kircher's fantastic profiles, Jessica Lucas's mind-melting features about online communities, Sydney Bradley and Amanda Perelli's tireless coverage of the business side of the creator world, Brandy Zadrozny's empathetic features about the internet, Sapna Maheshwari's detailed coverage of the influencer and ecommerce industry, and Rachel Greenspan's online expertise. There are so many more writers whom I could name. I'm grateful for all of the journalists whose work I've cited throughout this book.

This book is a personal history in many ways. Blogging, especially Tumblr, gave me everything I have. I am forever indebted to Kelly Bergin for introducing me to a platform that altered the course of my life. Thank you to every blogger who showed me the ropes, reblogged my posts, and let me crash their parties. And thank you to every early Tumblr staff member for creating the product that saved my life.

Cooper Fleishman gave me my first "official" byline at the *Daily Dot* and allowed me the freedom to chase stories and develop my beat at *Mic*. I would not be a reporter today without his encouragement.

Attending the first-ever XOXO Festival in Portland in 2012, organized by Andy Baio and Andy McMillan, inspired me to try to really make a go of writing about the internet.

I would not understand so much about the mechanics of virality without Neetzan Zimmerman, whom I was so lucky to have the opportunity to work for.

Thank you to the entire @taylorlorenz3.0 community for their feedback on everything from this book's name to its cover.

I wrote this book almost entirely from bed, as a medically vulnerable person (still!) trying to survive a deadly pandemic while being doxxed, stalked, harassed, and attacked by some of the worst corners of the internet. The people I met online who reached out and helped me through those dark days reaffirmed my faith in the internet and in technology.

I am so thankful for my parents and family members who, despite bearing the brunt of some of my worst online attacks, have always provided me with unwavering support. To the biggest reader I know, who supported me in every way possible during this whole endeavor, I could have never done this without you.

This book would not have come together without the hard work of so many at Simon & Schuster, including Brittany Adames, Emily Simonson, and Priscilla Painton in editorial; Stephen Bedford in marketing; Elizabeth Herman, Martha Langford, and Julia Prosser in publicity; Amanda Mulholland, Lauren Gomez, and Zoe Kaplan in managing editorial; Amy Medeiros, the production editor; Lexy East for interior design and desktop; Beth Maglione, Navorn Johnson, and Samantha Cohen in production; Mikaela Bielawski, who handled the ebook; copyeditor Rachelle Mandik; proofreaders Ashley Patrick and Vivian Reinert; art director Jackie Seow; Math Monahan and Emma Shaw in design; Tom Spain in audio; Ray Chokov and Nicole Moran in pre-press; Marie Florio and Mabel Taveras, who handled subrights; Lyndsay Brueggemann and Winona Lukito, who handled demand planning; and Simon & Schuster's publisher, Jonathan Karp, and associate publisher, Irene Kheradi. And thank you to Alan Dino Hebel and Ian Koviak of the Book Designers for the cover design.

// NOTES //

Introduction: The Social Ranking

2 **On a website with a pale:** socialrank, "News from Social Rank," Social Rank, July 16, 2006, http://web.archive.org/web/20060716102836/http://socialrank.wordpress.com/2006/04.

2 **"It all started with a meeting":** socialrank, "New York Social Elite Power Ranking (Ranking Period April 26, 2006–May 10, 2006)," Social Rank, May 5, 2006, http://web.archive.org/web/20060505235304/http://socialrank.wordpress.com/2006/04/24/new-york-social-elite-power-ranking-ranking-period-april-26-2006-may-10-2006.

3 **"All spots are up for grabs":** socialrank, "NY Social Elite Power Ranking (July 19–August 2, 2006)," Social Rank, August 19, 2006, http://web.archive.org/web/20060819140205/http://socialrank.wordpress.com/2006/07/19/ny-social-elite-power-ranking-july-19-august-7-2006.

3 **The comments section filled with insults:** Jessica Pressler, "How an Anonymous Gossip Website Changed New York Society Forever," *Town & Country*, August 10, 2016, https://www.townandcountrymag.com/society/a7307/socialite-rank.

3 **Within months:** Isaiah Wilner, "The Number-One Girl," *New York Magazine*, May 4, 2007, https://nymag.com/news/people/31555.

4 **When the blog derided one socialite:** Wilner, "Number-One Girl."

4 **Then, on February 8, 2007:** Danica Lo, "Mean Girls Get Camera-Shy," *New York Post*, February 8, 2007, https://nypost.com/2007/02/08/mean-girls-get-camera-shy.

4   **At first, Palermo received a warm reaction:** Pressler, "How an Anonymous Gossip Website Changed New York Society Forever."

5   **ran a cover story:** Wilner, "Number-One Girl."

5   **By 2016 she was reportedly:** Pressler, "How an Anonymous Gossip Website Changed New York Society Forever."

6   **When the internet first emerged:** Janna Quitney Anderson, *Imagining the Internet: Personalities, Predictions, Perspectives* (Rowman & Littlefield, 2005), 78.

Chapter 1: The Blogging Revolution

11 **The "web log" originated:** Jeffrey Rosen, "Your Blog or Mine?," *New York Times Magazine*, December 19, 2004, https://www.nytimes.com/2004/12/19/magazine/your-blog-or-mine.html.

12 **some of the first notable blogs:** Dianna Gunn, "The History of Blogging: From 1997 Until Now (With Pictures)," Themeisle blog, February 6, 2023, https://themeisle.com/blog/history-of-blogging/#gref.

12 **As the Bush-Gore contest intensified:** "Q&A with Josh Marshall," C-SPAN, January 12, 2012, https://www.c-span.org/video/?303536-1/qa-josh-marshall.

13 **Within months:** David Glenn, "The (Josh) Marshall Plan," *Columbia Journalism Review*, https://www.cjr.org/feature/the_josh_marshall_plan.php.

13 **The total number of blogs:** Rodd Zolkos, "First Word: Big Blog Bang, World of Opportunity," Business Insurance, https://www.businessinsurance.com/article/20061119/STORY/100020270?template=printart.

14 **They know that behind the curtain:** Andrew Sullivan, "The Blogging Revolution," *Wired*, May 2002, https://www.wired.com/2002/05/the-blogging-revolution.

14 **In his remarks, Lott:** "Senator Thurmond 100th Birthday," C-SPAN, December 5, 2002, https://www.c-span.org/video/?174100-1/senator-thurmond-100th-birthday.

14 **While the *Washington Post* and ABC News:** "Q&A with Josh Marshall," C-SPAN.

14 **Marshall assembled a broad argument:** Sean Flynn, "Men of the Year: Give This Man a Pulitzer," *GQ*, November 13, 2007, https://www.gq.com/story/men-of-the-year-josh-marshall-alberto-gonzalez.

14 **Lott took to TV:** Jim Rutenberg and Felicity Barringer, "DIVISIVE WORDS: ON THE RIGHT; Attack on Lott's Remarks Has Come from Variety of Voices on the Right," *New York Times*, December 17, 2002, https://www.nytimes.com/2002/12/17/us/divisive-words-right-attack-lott-s-remarks-has-come-variety-voices-right.html.

15 **On December 13, 2002:** John Podhoretz, "THE INTERNET'S FIRST SCALP," *New York Post*, December 13, 2002, https://nypost.com/2002/12/13/the-internets-first-scalp.

15 **Marshall started using BlogAds in 2003:** Simon Owens, "This Guy Invented Blog Advertising. Here's What He's up to Now," The Business of Content, May 13, 2019, https://medium.com/the-business-of-content/this-guy-invented-blog-advertising-heres-what-he-s-up-to-now-ef1650d98774.

15 **ad revenue stream had grown:** "Q&A with Josh Marshall."

16 **That same year, Perez Hilton:** Andrea Chang, "Turning a Blog into an Empire," *Los Angeles Times*, June 13, 2008, https://www.latimes.com/archives/la-xpm-2008-jun-13-fi-howimadeit13-story.html.

16 **"Hunter's unfiltered nightlife shots":** Kyle Munzenrieder, "Mark Hunter (AKA 'The Cobrasnake') Revisits His Early Aughts Heyday," *W*, March 24, 2021, https://www.wmagazine.com/life/the-cobrasnake-mark-hunter-party-photographs-interview.

16 **Bloggers were suddenly sitting:** Kerry Folan, "Dolce & Gabbana Has Totally Made Up with the Sartorialist, and They Have This Video to Prove It," *Racked*, June 1, 2012, https://www.racked.com/2012/6/1/7723037/scott-schuman-and-garance-dor-would-like-to-redefine-successful-media.

16   **In a shocking upset:** Christina Binkley, "Bloggers Join Fashion's Front Row," *Wall Street Journal,* October 2, 2009, https://www.wsj.com/articles/SB10001424052748704471504574445222739373290.

16   **"Bloggers have ascended":** Eric Wilson, "Bloggers Crash Fashion's Front Row," *New York Times,* December 24, 2009, https://www.nytimes.com/2009/12/27/fashion/27BLOGGERS.html.

16   **As blogs boomed:** "Newspapers Fact Sheet," Pew Research Center's Journalism Project, June 29, 2021, https://www.pewresearch.org/journalism/fact-sheet/newspapers.

17   **In testimony before Congress:** Testimony of David Simon. Before the relevant Senate Committee on Commerce, Science, and Transportation Subcommittee on Communications, Technology, and the Internet Hearing on the Future of Journalism. May 6, 2009. (David Simon, *Baltimore Sun,* 1982–85, Blown Deadline Productions, 95-09, Baltimore, MD), https://www.commerce.senate.gov/services/files/9392D321-43E8-4053-BDBB-466070864D5E

17   **Major publications:** Benjamin Carlson, "The Rise of the Professional Blogger," *Atlantic,* September 11, 2009, https://www.theatlantic.com/magazine/archive/2009/09/the-rise-of-the-professional-blogger/307696.

Chapter 2: The Mommy Bloggers

19   **Blogs were offering content:** Kathryn Jezer-Morton, "Did Moms Exist Before Social Media?" *New York Times,* April 16, 2020, https://www.nytimes.com/2020/04/16/parenting/mommy-influencers.html.

20   **"The early blogs were all about":** Jezer-Morton, "Did Moms Exist."

21   **By the end of the decade:** "Socializing and Shopping the Power of Power Moms Online," Nielsen, May 2009, https://www.nielsen.com/insights/2009/socializing-and-shopping-the-power-of-power-moms-online.

22 **"I don't possess the juggling skills":** Heather B. Armstrong, "Chchchch-Changes," *Dooce*, August 13, 2004, https://dooce.com/2004/08/13/chchchch-changes.

22 **Despite Armstrong's trepidation:** Chavie Lieber, "She Was the 'Queen of the Mommy Bloggers.' Then Her Life Fell Apart," *Vox*, April 25, 2019, https://www.vox.com/the-highlight/2019/4/25/18512620/dooce-heather-armstrong-depression-valedictorian-of-being-dead.

22 **"What made you important enough":** Lisa Belkin, "Queen of the Mommy Bloggers," *New York Times Magazine*, February 23, 2011, https://www.nytimes.com/2011/02/27/magazine/27armstrong-t.html.

22 **Between seeking a paycheck and shutting down:** Belkin, "Queen of the Mommy Bloggers."

23 **"because I realized I didn't need":** Lieber, "She Was the 'Queen.'"

23 **Meanwhile, consumer brands:** "Marketing to Women - Women Control 80% of Spending," TrendSight, April 10, 2014, https://web.archive.org/web/20140410211225/http://www.trendsight.com/content/view/40/204.

23 **If you could get mothers talking:** "P&G's Vocalpoint - Using Moms for W.O.M.," ICMR marketing case study, 2006, https://www.icmrindia.org/casestudies/catalogue/marketing/P%20and%20G%20Moms.htm.

23 **In reviewing the results:** Jack Neff, "P&G Provides Product Launchpad, a Buzz Network of Moms," *Ad Age*, March 20, 2006, https://adage.com/article/news/p-g-product-launchpad-a-buzz-network-moms/107290; Bittner, Bill. "BrainTrust Query: Does P&G's Tremor Have the Formula for Building 'The Buzz'?" *RetailWire*, December 5, 2006, https://retailwire.com/discussion/braintrust-query-does-p-and-gs-tremor-have-the-formula-for-building-the-buzz.

24 **By 2007, BlogHer placed ads:** Janis Mara, "As More Women Flock Online, BlogHer Caters to What They Want," *East Bay Times*, November 19, 2007, www.eastbaytimes.com/2007/11/19

/as-more-women-flock-online-blogher-caters-to-what-they-want.

25 **"Brand executives and women":** Larissa Faw, "Is Blogging Really a Way for Women to Earn a Living?" *Forbes*, April 25, 2012, https://www.forbes.com/sites/larissafaw/2012/04/25/is-blogging-really-a-way-for-women-to-earn-a-living-2.

25 **"It happened so fast":** Belkin, "Queen of the Mommy Bloggers."

25 **For Armstrong, it was already:** Lieber, "She Was the 'Queen.'"

26 **she couldn't stop picturing:** XOXO Festival (@xoxofest), "Heather Armstrong, Dooce - XOXO Festival (2015)," YouTube, October 26, 2015, https://www.youtube.com/watch?v=fe-7kHmArAs.

26 **Banana Republic eventually relented:** Lieber, "She Was the 'Queen.'"

26 **"It was used in the media":** "Kathryn Jezer-Morton | Substack," Substack.com, substack.com/profile/116857-kathryn-jezer-morton.

Chapter 3: The Friend Zone

30 **But the stock price plummeted:** George Mannes, "The Rise and Fall of theglobe.com," *TheStreet*, May 26, 1999, https://www.thestreet.com/technology/the-rise-and-fall-of-theglobecom-750716.

30 **In its first two years:** "COMPANY NEWS; YOUTHSTREAM TO ACQUIRE SIXDEGREES FOR $125 MILLION," *New York Times*, December 16, 1999, https://www.nytimes.com/1999/12/16/business/company-news-youthstream-to-acquire-sixdegrees-for-125-million.html.

30 **Friendster became the hottest property:** Julia Angwin, *Stealing MySpace* (New York: Random House, 2009), chap. 6, iBooks.

30 **By the summer:** Angwin, *Stealing MySpace*, chap. 7.

31 **it started as a brazen:** Angwin, *Stealing MySpace*, chap. 2.

31 **He promised her that:** Angwin, *Stealing MySpace*, chap. 7.

31 **The crowd made the site seem fresh:** Angwin, *Stealing MySpace*, chap. 13.

32 **"It was kinda like a video game":** Sabrina Rubin Erdely, "Kiki Kannibal: The Girl Who Played with Fire," *Rolling Stone*, April 15,

33 2011, https://www.rollingstone.com/culture/culture-news/kiki-kannibal-the-girl-who-played-with-fire-66507.

33 **And for many, fashion subcultures:** Sara Tardiff, "Myspace's Greatest Gift Was Teaching Us How to Live Online," *Input*, March 28, 2020, https://www.inverse.com/input/culture/myspace-taught-us-to-craft-our-internet-personas-our-personal-style-live-online.

33 **the most visited website:** Angwin, *Stealing MySpace*, chap. 26.

33 **And by the end of the summer:** David Kirkpatrick, *The Facebook Effect* (Simon & Schuster, 2010), chap. 2, iBooks.

33 **Facebook remained limited:** Kirkpatrick, *Facebook Effect*, prologue.

34 **the user base grew from 1 million:** Kirkpatrick, *Facebook Effect*, chap. 9.

34 **MySpace had no friend limit:** Kirkpatrick, *Facebook Effect*, prologue.

34 **He spoke of Facebook as:** Kirkpatrick, *Facebook Effect*, chap. 6; Mark Zuckerberg, "Bringing the World Together," Facebook, March 15, 2021, https://www.facebook.com/notes/393134628500376.

34 **MySpace's founders saw:** Angwin, *Stealing MySpace*, chap. 11.

35 **As Tila Tequila told one reporter:** Angwin, *Stealing MySpace*, chap. 19.

35 **"a generation of suburban misfits":** Sandra Song, "Hanna Beth Invented the Influencer," *Paper*, July 29, 2019, https://www.papermag.com/hanna-beth-influencer-2639373172.html.

35 **"We were basically influencers":** Tardiff, "Myspace's Greatest Gift."

36 **Model Christine Dolce:** Sean Percival, "MySpace Marketing: Must-Have Friends," Informit, March 4, 2009, https://www.informit.com/articles/article.aspx?p=1312832&seqNum=3.

36 **In her last major headline:** Kate Aurthur, "Tila Tequila's Descent into Nazism Is a Long Time Coming," BuzzFeed, November 22, 2016, https://www.buzzfeed.com/kateaurthur/tila-tequilas-descent-into-nazism-is-a-long-time-coming.

36 **Christopher Stone founded:** Adrian Chen, "StickyDrama's Christopher Stone Is a 'Sextortion' Expert in More Ways than One," *Gawker*, July 24, 2010, www.gawker.com/5595591/sticky dramas-christopher-stone-is-a-sextortion-expert-in-more-ways-than-one.

37 **Gaming was huge:** Dean Takahashi, "Zynga's CityVille Becomes the Biggest-Ever App on Facebook," VentureBeat, January 3, 2011, venturebeat.com/games/zyngas-cityville-becomes-the-biggest-ever-app-on-facebook/.

38 **it resold for $35 million:** Dominic Rushe, "Myspace Sold for $35m in Spectacular Fall from $12bn Heyday," *Guardian*, June 30, 2011, https://www.theguardian.com/technology/2011/jun/30/myspace-sold-35-million-news.

39 **This visibility led friends to share more:** Kirkpatrick, *Facebook Effect*, chap. 9.

39 **"Stalkers Rejoice!":** Pete Cashmore, "Stalkers Rejoice! Facebook Updates News Feed," Mashable, November 15, 2006, https://mashable.com/archive/stalkers-rejoice-facebook-updates-news-feed.

Chapter 4: The New Celebrity

42 **"Few celebrities":** John Leland, "URBAN FABLES; Once You've Seen Paris, Everything Is E = Mc² (Published 2003)," *New York Times*, November 23, 2003, www.nytimes.com/2003/11/23/style/urban-fables-once-you-ve-seen-paris-everything-is-e-mc2.html.

43 **"Paris really started that movement":** Paris Hilton, "The Real Story of Paris Hilton | This Is Paris Official Documentary," YouTube, September 13, 2020, www.youtube.com/watch?v=wOg0TY1jG3w.

43 **"She was so real":** "The Simple Life," Reality TV World, https://www.realitytvworld.com/realitytvdb/the-simple-life.

44 **When *The Simple Life* premiered:** "Second Episode of 'The

Simple Life' Draws Even Better Ratings," Reality TV World, https://www.realitytvworld.com/news/second-episode-of-the-simple-life-draws-even-better-ratings-2065.php.

44 **A couple of months before the show's release:** Zoe Guy, "Paris Hilton Says She Was Coerced into Making Sex Tape," *Vulture*, March 7, 2023, www.vulture.com/2023/03/paris-hilton-sex-tape-memoir.html.

44 **He reportedly made $10 million:** "Paris Hilton—I Never Made a Dime off My Sex Tape," *TMZ*, November 25, 2013, www.tmz.com/2013/11/25/paris-hilton-sex-tape-money-rick-salomon-1-night-in-paris/.

47 **Commenting relentlessly on someone's post:** Jason Tanz, "Internet Famous: Julia Allison and the Secrets of Self-Promotion," *Wired*, July 15, 2008, www.wired.com/2008/07/howto-allison/.

47 **When she showed up:** "Julia Allison Condom Fairy Halloween 2006," Raincoaster, January 6, 2011, https://raincoaster.com/2011/01/06/non-rebloggingnonsociety/julia-allison-condom-fairy-halloween-2006/#main.

47 **The post was vicious:** Chris Mohney, "Field Guide: Julia Allison," *Gawker*, November 1, 2006, https://www.gawker.com/211734/field-guide-julia-allison.

47 **One *Gawker* editor:** Jon Friedman, "*Gawker* Gets Respectable—and Remains Humorous," *MarketWatch*, August 15, 2007, https://www.marketwatch.com/story/correct-gawker-gasp-grows-up-and-gets-respectable.

48 **When he looked around:** Oliver Lindberg, "Interview with David Karp: The Rise and Rise of Tumblr," Lindberg Interviews, Medium, July 10, 2016, https://medium.com/the-lindberg-interviews/interview-with-david-karp-the-rise-and-rise-of-tumblr-ed51085140cd.

48 **"You could mix in a little":** Tim Shey, "Shey.net Reblog: On Re-Blogging," Shey.net, October 4, 2004, shey.net/reblog/2004/10/on_reblogging.html.

49 **"Anybody who's anybody":** Lauren Streib, "Webutante Ball: Julia Allison, Lockhart Steele, and More," *Daily Beast*, June 9, 2010, https://www.thedailybeast.com/articles/2010/06/09/chic-geeks.

50 **The party proved to be so popular:** "A NIGHT to ReMEMEber," Urlesque, i.kym-cdn.com/photos/images/newsfeed/000/619/064/af2.jpg.

51 *Page Six* **called her:** "No Pain No-Show," *New York Post*, April 30, 2009, https://pagesix.com/2009/04/30/no-pain-no-show.

51 **Unilever's chief marketing officer:** Glynnis MacNicol, "Julia Allison to Pen 'Lively' New Syndicated Column on Social Media," *Business Insider*, January 30, 2011, https://www.businessinsider.com/julia-allyson-to-pen-lively-new-syndicated-column-on-social-media-2011-1.

51 **Allison also signed a big deal with Sony:** "Sony Integrated Marketing Campaign Cuts Through the Retail Noise," Sony press release, August 18, 2009, https://www.sony.com/content/sony/en/en_us/SCA/company-news/press-releases/sony-electronics/2009/sony-integrated-marketing-campaign-cuts-through-the-retail-noise.html.

51 **One of them jeered:** Brian Moylan, "Julia Allison's Sony Commercials Offer a Window into Her Soul," *Gawker*, September 11, 2009, https://www.gawker.com/5357706/julia-allisons-sony-commercials-offer-a-window-into-her-soul.

52 *Fast Company* **ran a piece:** DJ Francis, "Sometimes Breasts Aren't Enough, Julia Allison," *Fast Company*, July 28, 2008, https://www.fastcompany.com/943818/sometimes-breasts-arent-enough-julia-allison.

52 *Radar* **magazine named her:** Paul Bradley Carr, "If You Can't Say Anything Nice, Then Kill Yourself Now," *Guardian*, January 28, 2009, www.theguardian.com/technology/2009/jan/28/not-safe-for-work-techcrunch-arrington.

53 **"If you follow JA now":** PrestoChango0804, "She's Just a 2020 Version of Julia Allison," Reddit, April 13, 2020, www.reddit.com/r/SmolBeanSnark/comments/g0tlni/comment/fnbz4a0.

53 **Allison, however, has not been blessed:** Erin Rodrigue, "31 Influencer Marketing Stats to Know in 2023," Hubspot Blog, January 16, 2023, blog.hubspot.com/marketing/influencer-marketing-stats.

53 **When Silicon Valley investors:** Kate Clark and Amir Efrati, "Andreessen Horowitz Wins Deal for Creator Economy Startup Stir at $100 Million Valuation," *Information*, February 10, 2021, https://archive.ph/o/zi2o8/https:/www.theinformation.com/articles/andreessen-horowitz-wins-deal-for-creator-economy-startup-stir-at-100-million-valuation.

54 **"[Allison] represented a moment":** Heather Gold (@heathr), "I saw what she did as a bellwether. She was not accepted at all by web culture. But what she did became commonplace. That has more to do with the straight men of the old web and investors than her," Twitter, June 3, 2018, 7:02 p.m., https://twitter.com/heathr/status/1003411706326839298.

Chapter 5: The Rise of YouTube

59 **In early 2005, a dating website:** Jason Koebler, "10 Years Ago Today, YouTube Launched as a Dating Website," *Vice*, April 23, 2015, https://www.vice.com/en/article/78xqjx/10-years-ago-today-youtube-launched-as-a-dating-website.

59 **"We should just be a site where":** Mark Bergen, *Like, Comment, Subscribe* (New York: Viking, 2022), 22.

60 **By June, YouTube launched a beta:** JR Raphael, "YouTube's Anniversary: How HOTorNOT Started It All," *PCWorld*, October 9, 2009, https://www.pcworld.com/article/520072/youtubes_anniversary_how_hotornot_started_it_all.html.

60 **YouTube gained 10,000 users:** Naomi Jane Gray, "Viacom and YouTube Open the KimonoParties Publicly File Redacted Copies of Summary Judgment Motions," Shades of Gray, March 19, 2010, https://www.shadesofgraylaw.com/2010/03/18/viacom-and-youtube-open-the-kimonoparties-publicly-file-redacted-copies-of-summary-judgment-motions.

60   **Playing to their strengths:** Chris Osterndorf, "A Timeline of the Lonely Island's Trail-Blazing Internet Comedy," *Daily Dot*, June 3, 2016, https://www.dailydot.com/upstream/andy-samberg-lonely-island-snl-popstar.

60   **YouTube was still relatively unknown:** Andrew Wallenstein and Todd Spangler, "'Lazy Sunday' Turns 10: 'SNL' Stars Recall How TV Invaded the Internet," *Variety*, December 18, 2015, https://variety.com/2015/tv/news/lazy-sunday-10th-anniversary-snl-1201657949.

61   **YouTube's total traffic went up 83 percent:** Bill Higgins, "Hollywood Flashback: 'SNL's' 'Lazy Sunday' Put YouTube on the Map in 2005," *Hollywood Reporter*, October 5, 2017, https://www.hollywoodreporter.com/business/digital/hollywood-flashback-snls-lazy-sunday-put-youtube-map-2005-1044829.

61   **"Lazy Sunday" became a blueprint:** Wallenstein and Spangler "'Lazy Sunday' Turns 10."

61   **After "Lazy Sunday," YouTube's growth:** The YouTube Team, "Your Features Have Arrived!" YouTube Official Blog, February 27, 2006, https://blog.youtube/news-and-events/your-features-have-arrived.

62   **Bree posted her first video:** lonelygirl15 (@lonelygirl15), "First Blog / Dorkiness Prevails," YouTube, June 16, 2006, www.youtube.com/watch?v=-goXKtd6cPo.

63   **"If we didn't do it, then someone else":** Elena Cresci, "Lonelygirl15: How One Mysterious Vlogger Changed the Internet," *Guardian*, June 16, 2016, www.theguardian.com/technology/2016/jun/16/lonelygirl15-bree-video-blog-youtube.

64   **Rose was wary at first:** Cresci, "Lonelygirl15."

65   **When Amanda Goodfried responded:** Richard Rushfield and Claire Hoffman, "Mystery Fuels Huge Popularity of Web's Lonelygirl15," *Los Angeles Times*, September 8, 2006, https://www.latimes.com/entertainment/la-et-lonelygirl15-story.html.

65   **His father, Tom Foremski:** "How the Secret Identity of Lonely Girl15 Was Found," *Silicon Valley Watcher*, September 12, 2006,

https://www.siliconvalleywatcher.com/how-the-secret-identity-of-lonelygirl15-was-found.

65 **Abu-Taleb continued to play Daniel:** Joshua Davis, "The Secret World of Lonelygirl," *Wired*, December 1, 2006, https://www.wired.com/2006/12/lonelygirl.

65 **"Anybody could do what we did":** Richard Rushfield and Claire Hoffman, "Lonelygirl15 Video Blog Is Brainchild of 3 Filmmakers," *Los Angeles Times*, September 13, 2006, www.latimes.com/archives/la-xpm-2006-sep-13-me-lonelygir13-story.html.

66 **Over the summer of 2006:** Gavin O'Malley, "YouTube Is the Fastest Growing Website," *Ad Age*, July 21, 2006, https://adage.com/article/digital/youtube-fastest-growing-website/110632.

66 **Just before Bree took the top spot:** John Sutherland, "Geriatric1927—A Review of His Work," *Guardian*, August 30, 2006, https://www.theguardian.com/g2/story/0,,1860779,.html.

66 **By mid-2006, Susan Wojcicki:** Peter Kafka, "How YouTube Swallowed the World," *Vox*, March 2, 2021, https://www.vox.com/recode/22308263/youtube-google-land-of-the-giants-podcast.

66 **She and Google leadership went back:** The Associated Press, "Google Buys YouTube for $1.65 Billion," NBC News, October 9, 2006, www.nbcnews.com/id/wbna15196982.

71 **"It started getting like 30 hits a day":** Alison Foreman, "The Legend of Keyboard Cat: How a Man and His Cat(s) Won the Internet Lottery," Mashable, October 24, 2020, https://mashable.com/article/keyboard-cat-history.

71 **"I was counting on not getting that":** Foreman, "Legend of Keyboard Cat."

73 **At the time, the concept of paying:** "Posters Reap Cash Rewards at Video-Sharing Site Revver," *USA Today*, September 13, 2007, http://usatoday30.usatoday.com/tech/webguide/internetlife/2007-09-13-revver_N.htm.

74 **And more so than just TV:** Berkeley Arts + Design (@Berkeley ArtsDesign), "Cultural Criticism in the Age of YouTube with George Strompolos, Rolla Selbak and Tiffany Shlain,"

YouTube, May 6, 2017, https://www.youtube.com/watch?v=I2DVKR266_4.

74 **"To me, the revolution"**: Berkeley Arts + Design, "Cultural Criticism."

75 **They included Lonelygirl15:** The YouTube Team, "YouTube Elevates Most Popular Users to Partners," YouTube Official Blog, May 3, 2007, https://blog.youtube/news-and-events/youtube-elevates-most-popular-users-to/.

75 **Throughout the rest of 2007:** YouTube, "History of Monetization at YouTube—YouTube5Year," sites.google.com/a/pressatgoogle.com/youtube5year/home/history-of-monetization-at-youtube.

76 **The highlight of Bahner's year:** Jogwheel (@Jogwheel), "Inside YouTube Live ! – My Trip To The Concert / Gathering," YouTube, December 10, 2008, https://www.youtube.com/watch?v=LhCwll6BOWE.

77 **"We needed more":** Paige Leskin, "YouTube Is Now a Money-Making Machine, but the Platform's Early Success Was Fueled by Group of 'Misfits' Who Wrote the Rulebook for Internet Fame," *Business Insider*, June 3, 2020, https://www.businessinsider.com/youtube-15-anniversary-early-creators-shaped-platform-viral-monetization-influencers-2020-6.

77 **Back in 2006, Next New Networks:** "Next New Networks," Wayback Machine, June 18, 2011, web.archive.org/web/20110619025411/http:/www.nextnewnetworks.com/team/.

77 **"You contribute your own videos":** Alexa Crawls, "Next New Networks," Web.archive.org, January 8, 2007, web.archive.org/web/20070108135328/www.nextnewnetworks.com/networks.html.

78 **"challenging the idea":** Brad Stone, "Internet Start-up Plans Video-Oriented Sites on Niche Topics," *New York Times*, March 8, 2007, https://www.nytimes.com/2007/03/08/technology/08iht-video.html.

78 **also inked a deal with Julia Allison:** Peter Kafka, "Want More Julia Allison? Next New Networks Has You Covered," *Business Insider*, October 1, 2008, https://www.businessinsider.com/2008/10/want-more-julia-allison-next-new-networks-has-you-covered; TMIweekly (@TMIweekly), YouTube channel, https://www.youtube.com/tmiweekly.

79 **The house, at 419 Grand:** Kevin Nalts, "Top YouTube Stars Convene 'the Station': A Modern Brat Pack & YouTube You-Topia? | Will Video for Food," Web.archive.org, March 10, 2014, web.archive.org/web/20140310231908/http:/willvideoforfood.com/2009/09/07/top-youtube-stars-convene-the-station-a-modern-brat-pack-and-youtube-youtopia.

79 **"a mecca for the YouTube creator-community":** Eriq Gardner, "Maker Studios Lawsuit: Inside the War for YouTube's Top Studio," *Hollywood Reporter*, October 24, 2013, www.hollywoodreporter.com/business/business-news/maker-studios-lawsuit-inside-war-650541.

79 **"Every YouTube star":** Gardner, "Maker Studios Lawsuit."

80 **Over the course of 2009:** Jim O'Neill, "eMarketer: Online video ad spend soars in 2009," Fierce Video, December 14, 2009. https://www.fiercevideo.com/online-video/emarketer-online-video-ad-spend-soars-2009

81 **Ezra Cooperstein, VP and director:** Brian Morrissey, "Carl's Jr. Takes Bite into YouTube World," *Ad Week*, June 1, 2009, https://www.adweek.com/performance-marketing/carls-jr-takes-bite-youtube-world-110867.

81 **"Presumably, the videos":** Benjamin Wayne, "YouTube Is Doomed," *Business Insider*. April 9, 2009, https://www.businessinsider.com/is-youtube-doomed-2009-4.

82 **Next New Networks changed its strategy:** Wayback Machine, "Next New Creators." Archive.org, 2020, web.archive.org/web/20180704174327/http:/www.nextnewnetworks.com/page/next-new-creators.

Chapter 6: Creators Break Through

85 **Previous YouTube meet-ups:** davidjr.com (@davidjrdotcom), "Youtube Meetup 777," YouTube, May 12, 2009, www.youtube.com/watch?v=sKwRWhFf5XQ.

86 **"We wanted to make VidCon reflect":** Jenni Powell, "Top 10 VidCon Moments: Music, Vlogging, and Being Awesome," Tubefilter, July 12, 2010, www.tubefilter.com/2010/07/12/top-10-vidcon-moments-music-vlogging-and-being-awesome.

86 **"I don't know what it was":** Natalie Jarvey, "VidCon at 10: How a "Thrown-Together" Event Gave Rise to the Influencer Era," *Hollywood Reporter*, July 13, 2019, www.hollywoodreporter.com/news/general-news/vidcon-at-10-how-a-thrown-together-event-gave-rise-influencer-era-1224136.

87 **At the height of the conference:** HallWoodProductions (@HallWoodProductions), "Vidcon – The Station – Eenie Meenie Bikini (Justin Bieber Parody) LIVE," YouTube, July 10, 2010, https://www.youtube.com/watch?v=TreshM2bUFE.

87 **Falon Fatemi, a business development strategist:** The Youtube Team, "Our Highlights from Vidcon," YouTube Official Blog, July 12, 2010, blog.youtube/news-and-events/our-highlights-from-vidcon.

88 **The Partner Grant Program distributed:** Jason Kincaid, "YouTube Announces Partner Grants Program, Support for 4K Video Resolution," *TechCrunch*, July 9, 2010, techcrunch.com/2010/07/09/youtube-partner-program-4k; Tom Pickett, "Celebrating our partners' success," YouTube Official Blog, December 22, 2010, https://youtube.googleblog.com/2010/12/celebrating-our-partners-success.html.

88 **By the end of 2010, YouTube had enrolled:** Tom Pickett, "Supercharging the "Next" phase in YouTube partner development," YouTube Official Blog, March 7, 2011, https://blog.youtube/news-and-events/supercharging-next-phase-in-youtube.

88 **Next New Networks' Gregory Brothers released:** Will Wei,

"The 'Bed Intruder Song' From Auto-Tune The News Cracks Billboard's Hot 100," *Business Insider*, August 20, 2010, https://www.businessinsider.com/the-bed-intruder-song-from-auto-tune-the-news-cracks-billboards-hot-100-2010-8.

88 **The resulting track:** "Billboard Hot 100," *Billboard*, January 2, 2013, https://www.billboard.com/charts/hot-100/.

88 **The song sold nearly 100,000 copies on iTunes:** schmoyoho (@schmoyoho), "BED INTRUDER SONG!!!" YouTube, July 31, 2010, http://www.youtube.com/watch?v=hMtZfW2z9dw.

88 **Nebraskan Lucas Cruikshank:** David Sarno, "Fred's YouTube Channel Is Programming for Kids by Kids," Web Scout blog, *Los Angeles Times*, June 24, 2008, https://web.archive.org/web/20080701072000/https://latimesblogs.latimes.com/webscout/2008/06/freds-youtube-c.html.

88 **Cruikshank was raking in:** David Sarno, "YouTube Sensation Fred to Guest Star on ICarly next Monday," *Technology* (blog), *Los Angeles Times*, February 10, 2009, https://www.latimes.com/archives/blogs/technology-blog/story/2009-02-09/youtube-sensation-fred-to-guest-star-on-icarly-next-monday.

88 **It was the first Hollywood production:** FRED (@Fred), "'Fred: The Movie' Official Clip – 'Fred Gets Advice From His Dad About Women,'" YouTube, August 17, 2010, https://www.youtube.com/watch?v=rVFJzN20jhQ.

89 **"There's a turn in the tide happening":** Marc Hustvedt, "My Damn Channel Tacks On $4.4M, Investors Bullish Again on Web Series," Tubefilter, August 9, 2010, https://www.tubefilter.com/2010/08/09/my-damn-channel-tacks-on-4-4m-investors-bullish-again-on-web-series.

89 **Maker Studios raised $1.5 million:** GRP partners is now known as Upfront Ventures.

89 **Comedy network My Damn Channel secured:** Hustvedt, "My Damn Channel."

89 **Machinima raised $9 million:** "Series B - Machinima - 2010-

06-15 - Crunchbase Funding Round Profile," Crunchbase, https://www.crunchbase.com/funding_round/machinima-series-b--86cf3865.

90 **"a meeting with Revlon"**: Berkeley Arts + Design (@Berkeley ArtsDesign), "Cultural Criticism in the Age of YouTube with George Strompolos, Rolla Selbak and Tiffany Shlain," YouTube, May 6, 2017, https://www.youtube.com/watch?v=I2DVKR266_4.

90 **"I blurted out all this cool stuff"**: Berkeley Arts + Design, "Cultural Criticism."

90 **"And they would say"**: Berkeley Arts + Design, "Cultural Criticism."

90 **"And I had one of those moments where"**: Berkeley Arts + Design, "Cultural Criticism."

91 **"So much innovation comes from this group"**: Andrew Wallenstein, "Media Bigs Flock to YouTube Power Players," *Variety*, November 22, 2012, variety.com/2012/digital/news/media-bigs-flock-to-youtube-power-players-1118062549.

92 **So in March 2011, YouTube announced**: Tom Pickett, "Supercharging the 'Next' phase in YouTube partner development," YouTube Official Blog, March 7, 2011, https://blog.youtube/news-and-events/supercharging-next-phase-in-youtube.

93 **a group of eight college students**: Reyhan Harmanci, "Young, Cool . . . and Into Online Influence Metrics?" BuzzFeed News, August 23, 2012, https://www.buzzfeednews.com/article/reyhan/young-cooland-into-online-influence-metrics.

95 **"It's hard to overstate how big the hype"**: Jeremy Blacklow, "The Biggest Celebrity at SXSW: Grumpy Cat," Yahoo Entertainment, March 12, 2013, http://omg.yahoo.com/blogs/celeb-news/biggest-celebrity-sxsw-grumpy-cat-161435765.html.

95 **"Forget Elon Musk or Al Gore"**: Brandon Griggs, "The Unlikely Star of SXSW: Grumpy Cat," CNN, March 10, 2013, https://www.cnn.com/2013/03/10/tech/web/grumpy-cat-sxsw/index.html.

96 **Her Instagram account is still dutifully:** Grumpy Cat (@real grumpycat), "Doors are boring. Send a message with Grumpy Cat fabric door coverings from DoorFoto! There's an option for every occasion and holiday! Get 25% off your order now with the promo code GRUMPY25," Instagram photo, September 17, 2021, https://www.instagram.com/p/CT8Q0_KMPI3.

96 **He later said that at that time about 40 percent:** Berkeley Arts + Design, "Cultural Criticism."

97 **reported $28.8 billion in ad revenue:** Andrew Hutchinson, "YouTube Generated $28.8 Billion in Ad Revenue in 2021, Fueling the Creator Economy," Social Media Today, February 2, 2022, https://www.socialmediatoday.com/news/youtube-generated-288-billion-in-ad-revenue-in-2021-fueling-the-creator/618208.

Chapter 7: Twitter Follows Back

102 **Williams, previously the founder of Blogger:** Nick Bilton, *Hatching Twitter* (New York: Portfolio/Penguin, 2013), "Just Setting Up My Twttr" chapter, iBooks.

102 **Like a blog, a stream of updates:** Bilton, *Hatching Twitter*, "Just Setting Up My Twttr" chapter.

103 **On March 21, 2006, Dorsey posted the first tweet:** Jack Dorsey (@jack), "just setting up my twttr," Twitter, March 21, 2006, twitter.com/jack/status/20?lang=en.

103 **"for staying in touch and keeping up":** "Twitter—A Whole World in Your Hands," Twitter, 2006, web.archive.org/web/20061021175824/twitter.com/faq.

103 **"Taking a walk to get a bus pass":** Jack Dorsey (@jack), "Taking a walk to get a bus pass. Warm outside.," Twitter, January 1, 2007, web.archive.org/web/20070104122256/twitter.com/jack.

103 **"Randi and I just lip dubbed":** Julia Allison (@juliaallison), "Randi and I just lip dubbed Like a Virgin in matching white suits. Hysterical :)," Twitter, April 26, 2008, https://web.archive.org/web/20080428194921/twitter.com/juliaallison.

103 **"Twitter is a bad, bad thing":** Andrew Kantor, "Twitter Is Just Too

Much Information - USATODAY.com," *USA Today*, January 8, 2009, web.archive.org/web/20090108043352/http:/www.usatoday.com/tech/columnist/andrewkantor/2007-04-05-twitter_N.htm.

103 **"Who really cares"**: Alex Beam, "Twittering with Excitement? Hardly," Boston.com, August 16, 2008, archive.boston.com/lifestyle/articles/2008/08/16/twittering_with_excitement_hardly/.

103 **"Each little update"**: Clive Thompson, "Brave New World of Digital Intimacy," *New York Times*, September 5, 2008, www.nytimes.com/2008/09/07/magazine/07awareness-t.html?searchResultPosition=7.

104 **Dorsey was sitting at his desk**: Bilton, *Hatching Twitter*, "Chaos Again" chapter.

104 **"I felt the earthquake"**: Bilton, *Hatching Twitter*, "Chaos Again" chapter.

104 **In this moment, Williams saw a vision**: Bilton, *Hatching Twitter*, "Chaos Again" chapter.

105 **no one at Twitter knew**: Bilton, *Hatching Twitter*, "Chaos Again" chapter.

105 **When he became CEO, he pulled rank**: Bilton, *Hatching Twitter*, "The *Time* 101" chapter.

106 **Twitter nixed the friends list**: @Biz, "Friends, Followers, and Notifications," Twitter Blog, July 19, 2007, web.archive.org/web/20080706131439/http:/blog.twitter.com/2007/07/friends-followers-and-notifications.html.

107 **The hashtag, for instance**: Zachary Seward, "The First-Ever Hashtag, @-Reply and Retweet, as Twitter Users Invented Them," Quartz, October 15, 2013, qz.com/135149/the-first-ever-hashtag-reply-and-retweet-as-twitter-users-invented-them.

107 **Open-source developer Chris Messina**: Tom Huddleston Jr., "This Twitter User 'Invented' the Hashtag in 2007—but the Company Thought It Was 'Too Nerdy,'" CNBC, January 9, 2020, www.cnbc.com/2020/01/09/how-chris-messina-got-twitter-to-use-the-hashtag.html.

107 **The co-founders suggested:** Bilton, *Hatching Twitter*, "Is Twitter Down?" chapter.

107 **But before they could get around to it, users:** Huddleston, "This Twitter User."

107 **Users were manually retweeting:** Seward, "The First-Ever Hashtag."

108 **declared it a watershed:** Daniel Teridman, "Ashton Outmaneuvers CNN to 1 Million on Twitter," CNET, April 17, 2009, www.cnet.com/tech/services-and-software/ashton-outmaneuvers-cnn-to-1-million-on-twitter/.

111 **"E-lebrities":** Eric Wilson and Cathy Horyn, "Fashion's New Order," *New York Times*, May 9, 2019, archive.nytimes.com/www.nytimes.com/interactive/2012/09/05/fashion/newyorkfashionweek-theneworder.html.

112 **"But their real impact, in 140 characters":** Wilson and Horyn, "Fashion's New Order."

114 **brought swift outrage from Facebook users:** David Kirkpatrick, *The Facebook Effect* (Simon & Schuster, 2010), chap. 9, iBooks.

114 **dissonance between being a public and a private platform:** Kirkpatrick, *Facebook Effect*, chap. 17.

Chapter 8: Tumblr Famous

116 **devoted to single, funny themes:** Christian Lander (@clander), Stuff White People Like blog, web.archive.org/web/20160309160905mp_/stuffwhitepeoplelike.com/; Joe Mande, *Look at This F*cking Hipster* (New York, St. Martin's Press, 2014).

116 **"You're on Tumblr and . . .":** Julia Carpenter, "A Complete History of 'F*** Yeah' Tumblrs, the Happiest Blogs on the Web," *Washington Post*, April 8, 2015, www.washingtonpost.com/news/the-intersect/wp/2015/04/08/a-complete-history-of-f-yeah-tumblrs-the-happiest-blogs-on-the-web.

116 **Its origins lay in the song:** Simmyisdead, "America – Fuck Yeah!" YouTube, February 28, 2007, web.archive.org/web/20080101223139/http://www.youtube.com/watch?v=sWS-FoXbjVI.

117 **"It's a clever and effective way":** Carpenter, "A Complete History of 'F*** Yeah' Tumblrs."

117 **"There were definitely these Tumblr celebrities":** Carpenter, "A Complete History of 'F*** Yeah' Tumblrs."

119 **all manner of memey Tumblrs:** Jessica Amason and Richard Blakely, *This Is Why You're Fat: Where Dreams Become Heart Attacks* (New York: HarperCollins, 2009); Luke Winkie, "Urban Outfitters Literature," *Dirt*, March 14, 2023, dirt.fyi/article/2023/03/urban-outfitters-literature.

119 **From 2009 to 2012, Tumblr grew:** M. G. Siegler, "Tumblr Is on Fire. Now over 6 Million Users, 1.5 Billion Pageviews a Month," *TechCrunch*, July 19, 2010, techcrunch.com/2010/07/19/tumblr-stats.

119 **top 50 websites in America:** Marcello Mari, "What Has Yahoo! Actually Acquired: A Snapshot of Tumblr in Q1 2013," GWI blog, May 21, 2013, blog.gwi.com/chart-of-the-day/what-has-yahoo-actually-acquired-a-snapshot-of-tumblr-in-q1-2013.

120 **BuzzFeed made most of its revenue through:** Andrew Marantz, *Antisocial* (New York: Viking, 2019), chap. 5, iBooks.

120 **It would attach, say, Verizon's logo:** Marantz, *Antisocial*, chap. 5.

121 **The caption read:** swiked, "guys please help me - is this dress white and gold, or blue and black? Me and my friends can't agree and we are freaking the fuck out," Whoa Wow Wow! Tumblr blog, February 27, 2015, https://web.archive.org/web/20150227014959/https://swiked.tumblr.com/post/112073818575/guys-please-help-me-is-this-dress-white-and.

122 **Holderness later told Digiday:** "Meet Cates Holderness, the BuzzFeed Employee behind #TheDress," *Digiday*, February 27, 2015, digiday.com/media/meet-cates-holderness-buzzfeed-employee-behind-thedress/.

122 **"There's a lot of debate on Tumblr about this":** Cates Holderness, "What Colors Are This Dress?" BuzzFeed, February 26, 2015, www.buzzfeed.com/catesish/help-am-i-going-insane-its-definitely-blue#.bbVbxNk0P.

122 **"More than anything it really underlines":** Adrianne Jeffries, "Why Did 'the Dress' Go Viral? We Asked Meme Traffic Expert Neetzan Zimmerman," *Vice*, February 26, 2015, www.vice.com/en/article/ae3aek/why-did-the-dress-go-viral-we-asked-meme-traffic-expert-neetzan-zimmerman.

122 **Soon, celebrities, including Taylor Swift:** "Taylor Swift Says The Dress Is Black and Blue," *Time*, February 27, 2015, https://time.com/3725450/dress-taylor-swift-black-blue-white-gold-celebrities.

123 **Caitlin McNeill never got rich:** Catherine Alford, "Blue and Black . . . Or White and Gold? Either Way, This Viral Dress is a Money-Maker," Penny Hoarder, https://www.thepennyhoarder.com/make-money/side-gigs/blue-and-black-or-white-and-gold-viral-dress/.

123 **the best thing that could happen:** Natalie Kitroeff, "Tumblr Posts Start Memes and Win Jobs," *Bits* (blog), *New York Times*, July 24, 2012, https://archive.nytimes.com/bits.blogs.nytimes.com/2012/07/24/tumblr-posts-start-memes-and-win-jobs.

123 **Nord created an account:** "Influencer Marketing and the Audiences You Dream of Reaching with James Nord, Founder of Fohr on Apple Podcasts," *Waves Social Podcast*, S02 E08, https://podcasts.apple.com/us/podcast/s02-e08-influencer-marketing-and-the-audiences-you/id1484888104?i=1000471650168.

124 **The first iteration was called Fohr Card:** Stuart Elliot, "Gant Rugger Devotes Ads to 'Bros' and Their Clothes," *New York Times*, January 22 2013, www.nytimes.com/2013/01/22/business/media/gant-rugger-devotes-ads-to-bros-and-their-clothes.html.

Chapter 9: Instagram's Influence

127 **One night in January 2011, Liz Eswein:** John McDermott, "How Liz Eswein Became the Most Influential Person on Instagram," *Digiday*, November 6, 2014, https://digiday.com/media/liz-eswein-became-influential-woman-instagram/.

127 **Her first photo, uploaded on January 11, 2011:** Liz Eswein (@newyorkcity), Instagram photo, January 14, 2011, www.instagram.com/p/BAvS7/.

128 **They were originally working on:** Sarah Frier, *No Filter* (New York: Simon & Schuster: 2020), 15–26.

129 **Upon launch, they imposed:** Frier, *No Filter*, 15–26.

129 **In fact, it grew faster:** Frier, *No Filter*, 26–27.

129 **The first post from a major celebrity:** Frier, *No Filter*, 35–36.

129 **Systrom and his small team took advice:** Frier, *No Filter*, 34.

130 **At one point, Instagram was forced to expand:** Frier, *No Filter*, 39-40.

131 **Early on, it became clear:** Frier, *No Filter*, 81. Eventually the company revamped the Popular page entirely, in favor of a personalized algorithmic "Explore" tab.

133 **Eswein won the race:** Red Bull (@redbull), "Congrats to @NewYorkCity—Liz won our race to 100k followers," November 22, 2011, https://www.instagram.com/p/VTe2Y/.

134 **Together, they would become:** Caroline Moss, "The First Family of Instagram," *New York Times*, December 31, 2014, www.nytimes.com/2015/01/01/style/the-first-family-of-instagram.html.

134 **One of her first deals was a Nike campaign:** Sarah Frier, *No Filter*, 82.

134 **when Nordstrom rolled out its new line:** Sarah Jones, "Nordstrom Employs New York Instagrammers to Personalize Fall Accessories," Retail Dive, https://www.retaildive.com/ex/mobilecommercedaily/nordstrom-employs-new-york-instagrammers-to-personalize-fall-accessories.

135 **"Instagram was not supposed to be about":** Sarah Frier, *No Filter*, 83.

136 **Amber Venz (now Amber Venz Box):** Dara Prant, "How Amber Venz Box Built RewardStyle and with It, a Billion-Dollar Business," Fashionista, July 13, 2018, https://fashionista.com/2018/07/amber-venz-box-rewardstyle-liketoknowit-career; "S2:E9 Amber Venz Box on Establishing LTK and Dominating Influencer Mar-

keting," *The Bottom Line* podcast, produced by Harvard Ventures, https://anchor.fm/harvard-ventures/episodes/S2-E9-Amber-Venz-Box-on-Establishing-LTK-and-Dominating-Influencer-Marketing-e16m165; Jo Piazza, "Confessions of an Influencer Whisperer," *Town & Country*, March 10, 2021, https://www.townandcountrymag.com/society/money-and-power/a35729363/amber-venz-box-rewardstyle-influencer-whisperer; "Billion Dollar Female Tech Founder Amber Venz Box," *Screw It Just Do It* podcast: #074, https://startupu.libsyn.com/074-billion-dollar-female-tech-founder-amber-venz-box.

139 **"That allowed a lot of people":** Sarah Frier, author interview.

139 **Instagram's December 2014 crackdown on spam:** Andy Cohen, (@Andy), "I lost 20k followers in the #InstagramPurge and I'm feeling cleaner than ever! Only real people, please! Twitter, next?" Twitter, December 18, 2014, https://twitter.com/Andy/status/545632944170487808.

140 **Instagram put out a statement:** Taylor Lorenz, "Chaos Ensues as Instagram Deletes Millions of Accounts," *Business Insider*, December 18, 2014, www.businessinsider.com/chaos-ensues-as-instagram-deletes-millions-of-accounts-2014-12.

140 **"When you engage in self-promotional behavior":** Lorenz, "Chaos Ensues."

141 **immediately became the most-retweeted:** Ellen DeGeneres (@EllenDeGeneres), "If only Bradley's arm was longer. Best photo ever. #oscars," Twitter, March 2, 2014, https://twitter.com/theellenshow/status/440322224407314432.

141 **the September issue of *Vogue*:** "The Instagirls: Joan Smalls, Cara Delevingne, Karlie Kloss, and More on the September Cover of *Vogue*," *Vogue*, August 18, 2014, https://www.vogue.com/article/supermodel-cover-september-2014.

Chapter 10: Vine Time

146 **Vine's promise was so enticing:** Peter Kafka and Mike Isaac, "Twitter Buys Vine, a Video Clip Company That Never

Launched," AllThingsD, October 9, 2012, https://allthingsd.com/20121009/twitter-buys-vine-a-video-clip-company-that-never-launched.

147 **"Our original beta had something"**: Casey Newton, "Why Vine Died," *Verge*, October 28, 2016, https://www.theverge.com/2016/10/28/13456208/why-vine-died-twitter-shutdown.

147 **"Almost immediately it became clear"**: Newton, "Why Vine Died."

147 **On January 23, 2013, Dick Costolo**: Dick Costolo (@dickc), "Steak tartare in six seconds. http://vine.co/v/bOIqn6rLeID via @dhof," Twitter, January 23, 2023, https://twitter.com/dickc/status/294124523714916353?.

147 **"Steak tartare in six"**: Olivia Waxman, "Watch the First Vine Ever Shared," *Time*, January 23, 2016, time.com/4187825/first-vine-ever/.

148 **Jarre generated the first viral video:** Jérôme Jarre, "Don't be afraid of Love!," Vine, April 16, 2013, vine.co/v/bFQ1KYbhWTQ.

148 **based on an early viral YouTube video:** adarkenedroom (@adarkenedroom), "Afraid of Technology," YouTube, April 8, 2008, www.youtube.com/watch?v=Fc1P-AEaEp8.

148 **In an interview with Ellen DeGeneres:** Milagros Menendez (@milagrosmenendez9758), "Vine Star Jerome Jarre on Ellen Show," YouTube, December 15, 2013, www.youtube.com/watch?v=QgIAF6a5rjQ.

148 **Johns looks at him and says:** Marcus Johns, "My dad is such a loser... #skateboard #old #fart #jk #lovehim #funny #comedy #loop #favthings #amazing #marcusjohns," Vine, April 13, 2013, vine.co/v/bFZwbm59bE3.

149 **He commemorated the moment:** Marcus Johns, "1 million followers! Thank you so much!," Vine, July 2, 2013, vine.co/v/haVr0rpWMwm.

149 **"The most talented comedian"**: team unruly, "Every Industry Should Tell Their Story in 6-Seconds," *Unruly*, June 11, 2014, unruly.co/blog/article/2014/06/11/every-industry-needs-learn-tell-story-6-seconds-says-vine-star-jerome-jarre/.

149 **put out a flier on Instagram:** Rudy Mancuso (@rudymancuso), "Central Park (Sheep Meadow), NYC on Saturday, the 27th. 3pm. Wear a Costume #Costumeparty," Instagram, July 25, 2013, www.instagram.com/p/cMjj_dnJmC/.

150 **Johns, Jarre, Megalis, and Mancuso:** Rudy Mancuso (@rudymancuso), Instagram photo, August 15, 2013, www.instagram.com/p/dCfmvynJuL/.

Chapter 11: A Tangle of Competitors, A New Era for Users

151 **Logan Paul appeared:** Matt Murray, "Vine Superstar Logan Paul Takes over TODAY Account," TODAY.com, November 18, 2013, www.today.com/popculture/vine-superstar-logan-paul-takes-over-today-account-2D11603772.

152 **"Videos are a very difficult medium":** Ellis Hamburger, "Instagram CEO Kevin Systrom: 'I'm Not Really One for Ritual. Life's More Interesting That Way,'" *Verge*, November 30, 2012, www.theverge.com/2012/11/30/3710112/instagram-ceo-kevin-systrom.

152 **"I think there's a big opportunity here":** Colleen Taylor, "The Hunt for an 'Instagram for Video' Is On, and Socialcam Wants the Crown," *TechCrunch*, April 11, 2012, techcrunch.com/2012/04/11/socialcam-instagram-for-video-michael-seibel-interview/.

152 **Socialcam, which launched in 2011:** Jason Kincaid, "After a Hot Start, Justin.tv Spins off Socialcam, Its 'Instagram for Video,'" *TechCrunch*, August 29, 2011, techcrunch.com/2011/08/29/after-a-hot-start-justin-tv-spins-off-socialcam-its-instagram-for-video/; Eric Eldon, "With Growth Accelerating, Socialcam's Mobile Video App Passes 3 Million Downloads," *TechCrunch*, December 7, 2011, techcrunch.com/2011/12/07/with-growth-accelerating-socialcams-mobile-video-app-passes-3-million-downloads/.

152 **over 13 million downloads:** Darrell Etherington, "Twitter Releases Vine for Android Smartphones as It Tops 13M Users," *TechCrunch*, June 3, 2013, https://techcrunch.com/2013/06/03

/twitter-releases-vine-for-android-smartphones-tops-13m-users.

152 **"What we did to photos":** Colleen Taylor, "Instagram Launches 15-Second Video Sharing Feature, with 13 Filters and Editing," *TechCrunch*, June 20, 2013, techcrunch.com/2013/06/20/facebook-instagram-video/.

153 **Snapchat users had shared:** Billy Gallagher, "You Know What's Cool? A Billion Snapchats: App Sees over 20 Million Photos Shared per Day, Releases on Android," *TechCrunch*, October 29, 2012, techcrunch.com/2012/10/29/billion-snapchats/.

153 **on a train ride through Mountain View:** Paige Leskin, "The Life of TikTok Head Alex Zhu, the Musical.Ly Cofounder in Charge of Gen Z's Beloved Video-Sharing App," *Business Insider*, November 24, 2019, https://www.businessinsider.com/tiktok-head-alex-zhu-musically-china-life-bio-2019-11.

153 **What if, Zhu thought, you could combine the two:** All That Matters (@BrandedAllThatMatters), "Social Music: Alex Zhu, Co-Founder and Co-CEO, Musical.ly, All That Matters 2017," YouTube, November 15, 2017, www.youtube.com/watch?v=Bxkw2c3qbxw.

154 **among a wave of early Snapchat creators:** "Mike Platco – the Shorty Awards," Shorty Awards, April 23, 2017, shortyawards.com/9th/mplatco.

155 **In her first video posted to the platform:** Ariel Martin (@babyariel), "BabyAriel's First Musical.ly Post | Baby Ariel," YouTube, August 22, 2015, www.youtube.com/watch?v=LNwqJNi80Rc.

157 **VYou allowed users:** Ryan Lawler, "Video Q&A Startup VYou Is Shutting Down Its Consumer Site to Focus on White-Label Opportunities," *TechCrunch*, March 29, 2013, techcrunch.com/2013/03/29/vyou-shut-down.

157 **"We may have reached a tipping point":** Ben Popper, "Twitter Reportedly Acquires Periscope, an App for Broadcasting Live Video," *Verge*, March 9, 2015, www.theverge.com/2015/3/9/8177519/twitter-reportedly-acquires-periscope.

158 **one of the top grossing social apps:** Ben Popper, "The Live-Streaming App Where Amateurs Get Paid to Chat, Eat, and Sleep on Camera," *Verge*, March 20, 2015, https://www.theverge.com/2015/3/20/8257141/younow-app-live-streaming-meerkat-amateur-video.

Chapter 12: Parallel Lines

161 **"They aren't broadcasting":** Ben Popper, "The Live-Streaming App Where Amateurs Get Paid to Chat, Eat, and Sleep on Camera," *Verge*, March 20, 2015, https://www.theverge.com/2015/3/20/8257141/younow-app-live-streaming-meerkat-amateur-video.

161 **Tayser Abuhamdeh began live-streaming:** "LIVE - Mr.Cashier Is Broadcasting on YouNow," YouNow, www.younow.com/Mr.Cashier. 2023.

162 **"I started to act like people were there":** Popper, "Live-Streaming App Where Amateurs Get Paid to Chat."

162 **amassed over 135,000 followers:** Popper, "Live-Streaming App Where Amateurs Get Paid to Chat."

163 **DreamWorks Animation fired the opening salvo:** George Szalai, "DreamWorks Animation to Acquire Online Teen Network AwesomenessTV," *Hollywood Reporter*, May 1, 2013, www.hollywoodreporter.com/movies/movie-news/dreamworks-animation-acquire-online-teen-450171/.

163 **the Walt Disney Company acquired Maker Studios:** Rachel Abrams, "Time Warner Leads $36 Mil Investment in Maker Studios," *Variety*, December 20, 2012, https://variety.com/2012/tv/news/time-warner-leads-36-mil-investment-in-maker-studios-1118063880; Ryan Lawler, "Maker Studios Raises Another $26 Million from Canal+, Singtel, and Others to Grow Its Business Overseas," *TechCrunch*, September 12, 2013, https://techcrunch.com/2013/09/12/maker-studios-26m.

163 **Otter Media, a web-video joint venture:** Peter Kafka, "AT&T & Chernin Buy Fullscreen, the Big YouTube Video Network," *Vox*,

September 22, 2014, https://www.vox.com/2014/9/22/11631150/att-chernin-buy-fullscreen-the-big-youtube-video-network.

163 **"These digital stars"**: Todd Spangler, "New Breed of Online Stars Rewrite the Rules of Fame," *Variety*, August 5, 2014, variety.com/2014/digital/news/shane-dawson-jenna-marbles-internet-fame-1201271428/.

164 **signed as many creators as possible:** Brooks Barnes, "Disney Buys Maker Studios, Video Supplier for YouTube," *New York Times*, March 25, 2014, www.nytimes.com/2014/03/25/business/media/disney-buys-maker-studios-video-supplier-for-youtube.html.

164 **Two years after the Disney acquisition:** Geoff Weiss, "Maker Studios Reportedly Slashing Its Creator Network of 'Thousands' to Just 300," Tubefilter, February 15, 2017, www.tubefilter.com/2017/02/15/maker-studios-slashing-network-to-300-creators/.

164 **the ex-boyfriend of video-game developer Zoë Quinn:** Casey Johnston, "Chat Logs Show How 4chan Users Created #GamerGate Controversy," *Ars Technica*, September 9, 2014, arstechnica.com/gaming/2014/09/new-chat-logs-show-how-4chan-users-pushed-gamergate-into-the-national-spotlight/.

165 **Gamergate was a watershed moment:** Joseph Bernstein, "The Disturbing Misogynist History of GamerGate's Goodwill Ambassadors," BuzzFeed News, October 30, 2014, www.buzzfeednews.com/article/josephbernstein/the-disturbing-misogynist-history-of-gamergates-g#.oxZjNaYNK8.

165 **"young reactionaries to MAGA":** Joan Donovan, Emily Dreyfuss, and Brian Friedberg, *Meme Wars* (New York: Bloomsbury, 2022), 305.

165 **central to Donald Trump's election:** Mike Snider, "Steve Bannon Learned to Harness Troll Army from 'World of Warcraft,'" *USA Today*, July 18, 2017, https://www.usatoday.com/story/tech/talkingtech/2017/07/18/steve-bannon-learned-harness-troll-army-world-warcraft/489713001/.

166 **Gary Vee would get coffee with him:** Gary Vaynerchuk (@Gary Vee), "How Gary Vaynerchuk Met Jerome Jarre," April 2, 2015, www.youtube.com/watch?v=0qiLDGvJsj0.

166 **"It took him all of seven minutes":** Nick Bilton, "Jerome Jarre: The Making of a Vine Celebrity," *New York Times*, January 28, 2015, www.nytimes.com/2015/01/29/style/jerome-jarre-the-making-of-a-vine-celebrity.html?_r=0.

166 **Grier was able to command $25,000:** Christopher Glazek, "The Weird World of Internet Fame—Jerome Jarre, the Vine Entrepreneur," *New York Magazine*, April 18, 2014, nymag.com/news/media/internet-fame/jerome-jarre-2014-4.

167 **for computer giant HP:** Zach King, "I love aquarium screensavers. #BendTheRules @HP," Vine, August 20, 2014, vine.co/v/MLzT3utLh7B.

167 **essentially a supercut of Vines produced by the creators:** Jeff Beer, "Vine Stars Combine Their Powers for This HP TV Commercial," *Fast Company*, August 11, 2014, www.fastcompany.com/3034241/this-hp-tv-commercial-is-made-completely-out-of-vines.

167 **Logan Paul traveled across the United States:** Jeff Beer, "Vine Star Logan Paul Brings His Six-Second Creativity to New Hanes Campaign," *Fast Company*, July 20, 2014, www.fastcompany.com/3033265/vine-star-logan-paul-brings-his-six-second-creativity-to-new-hanes-campaign.

167 **Niche was facilitating so many deals:** Alyson Shontell, "Twitter Buys Niche, an Ad Network for Vine Stars, for about $50 Million in Cash and Stock," *Business Insider*, February 11, 2015, www.businessinsider.com/twitter-buys-niche-an-ad-network-for-vine-stars-2015-2.

168 **was making close to $75,000 a year:** Peter Kafka, "Advertisers Want Internet Stars, and Niche Wants to Connect Them," *Vox*, May 15, 2014, www.vox.com/2014/5/15/11626908/advertisers-want-internet-stars-and-niche-wants-to-connect-them.

170 **garnered over 2.7 million views:** Vine Boys Norway (@vine

boysnorway3209), "Funny Moments Magcon Boys Pt.1," YouTube, March 9, 2014, www.youtube.com/watch?v=9NpcK0oYoPY.

170 **A compilation of Magcon-boy Vines:** aethina (@cloudylrh), "Compilation MAGCON Best Vines," YouTube, May 2, 2014, www.youtube.com/watch?v=fsGP7JdE1xY.

171 **"But the internet stars of Magcon":** Megan Willett-Wei, "After a Major Convention Announced Its Comeback, Drama Started Brewing between a Bunch of Vine Stars," *Business Insider*, September 15, 2015, www.businessinsider.com/is-magcon-coming-back-2015-9.

172 **DigiTour itself expanded aggressively:** Peter Kafka, "Ryan Seacrest and Conde Nast Parent Invest in a YouTube Concert Tour," *Vox*, May 15, 2014, https://www.vox.com/2014/5/15/11626912/ryan-seacrest-and-conde-nast-parent-invest-in-a-youtube-concert-tour.

173 **When he realized he couldn't have her:** Chris Stokel-Walker, "What the Murder of Christina Grimmie by a Fan Tells Us about YouTube Influencer Culture," *Time*, May 3, 2019, time.com/5581981/youtube-christina-grimmie-influencer/.

Chapter 13: Counting Seconds

175 **There was also a D.C.-based "blog house":** Ashley Parker, "Washington Doesn't Sleep Here," *New York Times*, March 9, 2008, www.nytimes.com/2008/03/09/fashion/09bloghouse.html.

178 **Cabalona, himself a recent college graduate:** Mashable, "Just doing some rearranging around the office... #vine," Vine, January 25, 2015, vine.co/v/b5QFFdnqgH9.

179 **a dedicated "Vine studio":** Mashable, "Inside the Mashable Vine Studio," Vine, August 13, 2013, vine.co/v/hMMWat5JLhF.

179 **introduced web profiles and vanity URLs:** Nicole Lee, "Vine Introduces Web Profiles, Lets You Snag Vanity URL," *Engadget*, December 20, 2013, www.engadget.com/2013-12-20-vine-web-profiles.html.

181 **The company launched the campaign:** Viners, "I'm @twitter

|IG|snap: thegabbieshow. Tune in to my own special VidCon channel this weekend," Vine, July 23, 2015, vine.co/v/egawZaQ2DvI.

182 **YouTube spent millions:** Sam Gutelle, "YouTube's Next Creator-Focused TV, Print Ads Will Feature Lilly Singh, Tyler Oakley," Tubefilter, September 8, 2015, www.tubefilter.com/2015/09/08/youtubes-next-creator-focused-ads-will-feature-lilly-singh-tyler-oakley/.

182 **"When it comes to promoting YouTube":** Mike Shields, "YouTube Touts Collective Digital Studio's 'Video Game High School' in New Ad Campaign," *Wall Street Journal*, September 9, 2014, www.wsj.com/amp/articles/youtube-touts-collective-digital-studios-video-game-high-school-in-new-ad-campaign-1410287826.

183 **to meet with First Lady:** Madison Malone Kircher, "The Most Popular Female Vine Star Just Got to Meet Michelle Obama," *Business Insider*, October 19, 2015, www.businessinsider.com/vine-stars-take-selfie-with-michelle-obama-2015-10.

185 **Bachelor tells a group of men and women:** Cody Ko (@CodyKo), "It's Gotten Worse . . ." YouTube, April 27, 2017, www.youtube.com/watch?v=X3vgL5IFWuU.

187 **didn't turn a profit until:** David Goldman, "Twitter Is Losing Customers and Its Stock Is Falling," CNN Money, February 10, 2016, https://money.cnn.com/2016/02/10/technology/twitter-stock-users/index.html.

189 **"Twitter's growth has stagnated":** Goldman, "Twitter Is Losing Customers."

Chapter 14: The Shuffle

192 **He was one of the last:** Logan Paul (@loganpaulvlogs), "very bad news.," YouTube, March 21, 2017, www.youtube.com/watch?v=GgL0Rh9O3V8.

192 **"It's a much better, direct way":** Taylor Lorenz, "Months before Vine's Demise, Its Biggest Stars Plotted Their Escape," Mic,

October 28, 2016, www.mic.com/articles/157945/months-before-vine-s-demise-its-biggest-stars-plotted-their-escape.

193 **"as far as the fans go":** Lorenz, "Months before Vine's Demise."

193 **Facebook videos every day:** Josh Althuser, "Facebook Will Be All Video in 5 Years: Here Are 4 Figures to Prove It," *Social Media Today*, July 15, 2016, www.socialmediatoday.com/marketing/facebook-will-be-all-video-5-years-here-are-4-figures-prove-it.

194 **Meanwhile, user-posted videos on Facebook:** Shannon Tien, "Top 5 Social Media Trends in 2019 (and How Brands Should Adapt)," Hootsuite Social Media Management, May 27, 2019, blog.hootsuite.com/social-media-trends/.

194 **"video will look like as big of a shift":** Maya Kosoff, "Facebook Exaggerated Its Video-View Metrics for Two Years," *Vanity Fair*, September 23, 2016, www.vanityfair.com/news/2016/09/facebook-exaggerated-its-video-view-metrics-for-two-years.

194 **"It's not the kind of traditional video":** Jessica Guynn, "Mark Zuckerberg Talks up Facebook's 'Video First' Strategy," *USA Today*, November 2, 2016, www.usatoday.com/story/tech/news/2016/11/02/mark-zuckerberg-talks-facebook-video-first/93206596/.

194 **"more people tuned in":** Brendan Klinkenberg, "This Exploding Watermelon Was Facebook Live's Biggest Hit to Date," BuzzFeed News, April 8, 2016, www.buzzfeednews.com/article/brendanklinkenberg/this-exploding-watermelon-was-facebook-lives-biggest-hit-to.

194 **also known as "Chewbacca Mom":** Candace Payne, "It's the Simple Joys in Life....," Facebook video, May 19, 2016, www.facebook.com/candaceSpayne/videos/10209653193067040/.

196 **As part of those efforts:** Deepa Seetharaman and Steven Perlberg, "Facebook to Pay Internet Stars for Live Video," *Wall Street Journal*, July 19, 2016, www.wsj.com/articles/facebook-to-pay-internet-stars-for-live-video-1468920602.

196 **His content traveled so far in the News Feed:** Seetharaman and Perlberg, "Facebook to Pay Internet Stars."

// NOTES // 331

196 **he did splits in famous locations:** Logan Paul, "SPLITTING THE WORLD! 🌍 #OlympicEdition 🦵 CREATE YOUR OWN SPLITS EMOJI WITH MY NEW APP! ➔ LINK HERE ➔ http://appstore.com/SplitMoji," Facebook video, August 8, 2016, www.facebook.com/LoganPaul/videos/538113089720206/.

196 **allowing some creators to place fifteen-second ads:** Daisuke Wakabayashi, "Why Some Online Video Stars Opt for Facebook over YouTube," *New York Times*, July 9, 2017, www.nytimes.com/2017/07/09/technology/facebook-video-stars.html.

198 **His effectiveness and addictiveness:** Jon Caramanica, "For DJ Khaled, Snapchat Is a Major Key to Success," *New York Times*, December 21, 2015, www.nytimes.com/2015/12/22/arts/music/for-dj-khaled-snapchat-is-a-major-key-to-success.html.

199 **"The disenchantment of Snapchat's top":** Katie Notopoulos, "Snapchat Use Is down 34% among Top Influencers," BuzzFeed News, October 3, 2017, www.buzzfeednews.com/article/katienotopoulos/snapchat-use-is-down-34-among-top-influencers.

200 **viral inspirational hustle porn:** Taylor Lorenz, "LinkedIn Bro Poetry Pretty Much Sums Up 2017," *Daily Beast*, December 8, 2017, https://www.thedailybeast.com/linkedin-bro-poetry-pretty-much-sums-up-2017.

201 **"I love Instagram":** Alex Kantrowitz, "Frustrated Snap Social Influencers Leaving for Rival Platforms," BuzzFeed News, March 2, 2017, www.buzzfeednews.com/article/alexkantrowitz/frustrated-snap-social-influencers-leaving-for-rival-platfor#.vob5Je404.

Chapter 15: The Winners

205 **"Cheerleader" shot to number 1:** Gary Trust, "OMI's 'Cheerleader' Leaps to No. 1 on Hot 100," *Billboard*, July 13, 2015, www.billboard.com/pro/omi-cheerleader-hot-100.

206 **Off the back of that success:** Meg Graham, "Don't Judge Challenge: Why Teens Are Getting 'Ugly' for Social Media," *Chicago*

Tribune, July 6, 2015, www.chicagotribune.com/business/blue-sky/ct-dont-judge-challenge-bsi-20150706-story.html.

206 **Within just four days:** Todd Spangler, "Musical.ly's Live.ly Is Now Bigger than Twitter's Periscope on IOS (Study)," *Variety*, September 30, 2016, variety.com/2016/digital/news/musically-lively-bigger-than-periscope-1201875105.

206 **Musical.ly's VidCon 2016 success:** "Short Video Service Musical.ly Is Merging into Sister App TikTok," *TechCrunch*, March 2, 2017, techcrunch.com/2018/08/02/musically-tiktok.

207 **"Musical.ly has taken the art of lip-syncing":** Dan Rys, "Fresh off a Big Funding Round, Musical.ly Signs Its First Major Label Deal with Warner Music," *Billboard*, June 29, 2016, www.billboard.com/pro/warner-music-group-deal-musical-ly.

207 **ByteDance redesigned Flipagram:** Aisha Malik, "ByteDance Accused of Scraping Content from Instagram, Snapchat," *TechCrunch*, techcrunch.com/2022/04/04/tiktok-owner-bytedance-reportedly-scraped-content-from-instagram-snapchat-posted-flipagram.

207 **sold to ByteDance for $860 million:** Jon Russell and Katie Roof, "China's Bytedance Is Buying Musical.ly in a Deal Worth $800M-$1B," *TechCrunch*, November 9, 2017, techcrunch.com/2017/11/09/chinas-toutiao-is-buying-musical-ly-in-a-deal-worth-800m-1b.

208 **YouTube gave $100 million:** Marc Hustvedt, "YouTube Reveals Original Channels," Tubefilter, October 29, 2011, www.tubefilter.com/2011/10/28/youtube-original-channels/.

208 **the company also introduced Creator Clubs:** YouTube, "YouTube Creator Clubs," sites.google.com/site/ytcreatorclubs/?pli=1.

209 **YouTube announced Google Preferred:** Andrew Wallenstein, "YouTube Unveils Google Preferred at NewFronts Event," *Variety*, April 30, 2014, https://variety.com/2014/digital/news/youtube-unveils-google-preferred-at-newfronts-event-1201168888/; Maria Yagoda, "You're About to See a Lot More of YouTube Creators Tyler Oakley and Lilly Singh," *People*, September 8, 2015,

https://web.archive.org/web/20170811080434/https://people.com/celebrity/lilly-singh-and-tyler-oakley-featured-youtube-creators-new-youtube-ad-campaign/.

209 **Leadership's embrace of watch time:** Mark Bergen, *Like, Comment, Subscribe* (New York: Viking, 2022), 154.

209 **After uploading videos:** Casey Neistat (@casey), "The Pressure of Being a YouTuber," YouTube, May 28, 2018, www.youtube.com/watch?v=G38ixvYVNyM.

210 **David Dobrik, who had also hopped over:** Geoff Weiss, "David Dobrik Taking Brief Vlogging Break so That His 420th Video Falls on 4/20," Tubefilter, April 5, 2018, www.tubefilter.com/2018/04/05/david-dobrik-brief-vlogging-break-420/.

211 **"YouTube please do something about":** Issa (@twaimz), "@youtube please do something about sam pepper because i'm actually sick to my stomach and that should not be allowed on the internet," Twitter, November 30, 2015, twitter.com/twaimz/status/671246139610365952.

211 **brands were spending more than $255 million:** Sarah Frier and Matthew Townsend, "Bloomberg - Are You a Robot?" *Bloomberg*, August 5, 2016, www.bloomberg.com/news/articles/2016-08-05/ftc-to-crack-down-on-paid-celebrity-posts-that-aren-t-clear-ads.

212 **Fittingly, in 2016, when Jake Paul:** Brie Hiramine, "What Is Team 10? Jake Paul's Social Media Incubator House Debunked," J-14, September 27, 2018, https://www.j-14.com/posts/what-is-team-10-137260; Jake Paul (@jakepaul), "We are officially Team 10 ⑩ @Team10official ⑩ follow us to keep up with the squad ♥," Twitter, August 6, 2016, twitter.com/jakepaul/status/762006604476583936.

212 **So-called Instagram museums:** Clare Lanaux, "The Museum of Ice Cream Pop-Up Is Now Open in Los Angeles," The Points Guy, April 30, 2017, https://thepointsguy.com/2017/04/museum-of-ice-cream-los-angeles.

212 **Bright, colorful walls:** Michelle Rae Uy, "Los Angeles' 12 Most Instagram-Worthy Walls," Fodor's Travel Guide, December 21, 2017, https://www.fodors.com/news/photos/los-angeles-12-most-instagram-worthy-walls.

212 **avocado toast:** Jasmine Vaughn-Hall, "These Yummy Avocado Toast Captions Are All You Avo Wanted For Your Pics," Elite Daily, July 25, 2018, https://www.elitedaily.com/p/these-avocado-toast-instagram-captions-are-all-you-avo-wanted-9879019.

212 **Millennial-pink everything:** Maura Judkis, "Millennial Pink Took over Your Instagram Feed. Now It's Coming for Your Food," *Washington Post*, October 23, 2021, https://www.washingtonpost.com/news/food/wp/2017/08/10/millennial-pink-took-over-your-instagram-feed-now-its-coming-for-your-food.

212 **This was the content that became synonymous with:** Zachary Carlsen, "16 Latte Art Pros You Need To Follow On Instagram Right Now," Sprudge, February 16, 2023, https://sprudge.com/16-instagram-latte-art-feeds-you-need-to-follow-61430.html.

212 **made thousands of dollars:** Taylor Lorenz, "Custom Photo Filters Are the New Instagram Gold Mine," *Atlantic*, November 13, 2018, www.theatlantic.com/technology/archive/2018/11/influencers-are-now-monetizing-custom-photo-filters/575686/.

212 **"Instagram husband":** @TheMysteryHour, "Instagram Husband," YouTube, December 8, 2015, www.youtube.com/watch?v=fFzKi-o4rHw.

213 **A Taco Bell ad released in the fall:** Qiyu Liu (@QiyuLiu), "Taco Bell Instagram Boyfriend Commercial," YouTube, December 1, 2018, www.youtube.com/watch?v=cYaz6LVnKIk.

Chapter 16: Peak Instagram

215 ***Bloomberg* reported that DJ Khaled:** Sarah Frier and Matthew Townsend, "FTC to Crack Down on Paid Celebrity Posts That Aren't Clear Ads," *Bloomberg*, August 5, 2016, https://www

.bloomberg.com/news/articles/2016-08-05/ftc-to-crack-down-on-paid-celebrity-posts-that-aren-t-clear-ads.

216 **Kylie launched a beauty and skincare:** Emma Akbareian, "Kylie Jenner Lip Filler Confession Leads to 70% Increase in Enquiries for the Procedure," *Independent*, May 7, 2015, www.independent.co.uk/life-style/fashion/news/kylie-jenner-lip-filler-confession-leads-to-70-rise-in-enquiries-for-the-procedure-10232716.html.

216 **When she tweeted about not using Snapchat:** Kaya Yurieff, "Snapchat Stock Loses $1.3 Billion after Kylie Jenner Tweet," CNN Money, February 22, 2018, money.cnn.com/2018/02/22/technology/snapchat-update-kylie-jenner/index.html.

217 **"For a lot of years it was really":** Frier and Townsend, "FTC to Crack Down."

217 **Instagram itself introduced a new feature:** Rich McCormick, "Instagram Tries to Beat Secret Celebrity #Sponcon with New Label," *Verge*, June 14, 2017, www.theverge.com/2017/6/14/15799024/instagram-sponsored-posts-label.

218 **"[Influencers are] eager to assert the legitimacy":** Paul Pastore, "Influencers Are Upset about Being Stigmatized for Sponsored Content," CR Fashion Book, May 3, 2017, crfashionbook.com/celebrity-a9600107-influencers-ftc-sponsored-content-fatigue/.

220 **"More than most companies":** Katie Richards, "Nordstrom Is a Unicorn: How the Department Store Finds Success with Influencer Collabs," *Glossy*, June 3, 2019, www.glossy.co/fashion/nordstrom-is-a-unicorn-how-the-department-store-finds-success-with-influencer-collabs/.

221 **4 out of every 5 referrals:** Rachel Strugatz, "Digital Download: The Power of Influencer Referrals," *WWD*, September 19, 2017, wwd.com/feature/influencers-chriselle-lim-man-repeller-leandra-medine-reward-style-drive-traffic-and-sales-10994073/.

222 **Cassie Freeman, a lifestyle creator:** Stephanie Merry, "Popular Dallas Blogger Fashions Chic Summer Collection with Nordstrom - CultureMap Dallas," Dallas Culture Map, May 31, 2018,

dallas.culturemap.com/news/innovation/05-31-18-blogger-hi-sugarplum-gibson-nordstrom-summer-collection/#slide=0.

222 **"I remember the day Suzie":** Merry, "Popular Dallas Blogger."

223 **Over half a decade before:** John Coogan (@JohnCooganPlus), "Why Billionaires LOVE MrBeast," YouTube, January 17, 2022, www.youtube.com/watch?v=uMr-9MfDOI0.

223 **initial crop of talent got married:** Alexa Tietjen, "DIGITAL DOWNLOAD: How DBA Fosters Successful Influencer Brands — Offline," *WWD*, August 9, 2019, wwd.com/feature/digital-brand-architects-influencer-brands-gal-meets-glam-1203235581/.

223 **founded a podcast network called Dear Media:** Taylor Lorenz, "How Dear Media Reinvented Internet Celebrity," *Washington Post*, January 2, 2023, www.washingtonpost.com/technology/2022/12/29/dear-media-women-podcasts.

224 **"It looked sponsored":** Taylor Lorenz, "Influencers Are Faking Brand Deals," *Atlantic*, December 18, 2018, www.theatlantic.com/technology/archive/2018/12/influencers-are-faking-brand-deals/578401.

225 **Monica Ahanonu, an illustrator:** Lorenz, "Influencers Are Faking Brand Deals."

226 **failed to pay the content creators:** Taylor Lorenz, "When a Sponsored Facebook Post Doesn't Pay Off," *Atlantic*, December 26, 2018, www.theatlantic.com/technology/archive/2018/12/massive-influencer-management-platform-has-been-stiffing-people-payments/578767/.

Chapter 17: The Adpocalypse

229 **a story that rocked the internet:** Rolfe Winkler, et al., "Disney Severs Ties with YouTube Star PewDiePie after Anti-Semitic Posts," *Wall Street Journal*, February 14, 2017, www.wsj.com/articles/disney-severs-ties-with-youtube-star-pewdiepie-after-anti-semitic-posts-1487034533.

230 **Months after the PewDiePie story went viral:** "DaddyOFive

Parents Lose Custody 'over YouTube Pranks,'" *BBC News*, May 2, 2017, www.bbc.com/news/technology-39783670.

231 **The Martins ceased creating content:** KC Baker, "DaddyOFive YouTube Parents Lose Custody of Two Kids," *People*, May 3, 2017, people.com/crime/controversial-daddyofive-youtube-parents-lose-custody-of-2-children-featured-in-prank-videos.

231 **did little to break the prank cycle:** Amelia Tait, "'It's Just a Prank, Bro': Inside YouTube's Most Twisted Genre," *New Statesman*, April 21, 2017, https://www.newstatesman.com/science-tech/2017/04/its-just-prank-bro-inside-youtube-s-most-twisted-genre.

232 **The content of some of PewDiePie's videos:** Aja Romano, "YouTube's Most Popular User Amplified Anti-Semitic Rhetoric. Again." *Vox*, December 13, 2018, www.vox.com/2018/12/13/18136253/pewdiepie-vs-tseries-links-to-white-supremacist-alt-right-redpill.

232 **The London *Times* discovered:** Alexi Mostrous, Head of Investigations, "Big Brands Fund Terror through Online Adverts," *Times*, February 9, 2017, www.thetimes.co.uk/article/big-brands-fund-terror-knnxfgb98.

233 **AT&T and Johnson & Johnson:** Sapna Maheshwari and Daisuke Wakabayashi, "AT&T and Johnson & Johnson Pull Ads From YouTube," *New York Times*, March 23, 2017, https://www.nytimes.com/2017/03/22/business/atampt-and-johnson-amp-johnson-pull-ads-from-youtube-amid-hate-speech-concerns.html.

233 **Phil Smith, director general of Isba:** Jamie Grierson, "Google Summoned by Ministers as Government Pulls Ads over Extremist Content," *Guardian*, March 17, 2017, www.theguardian.com/technology/2017/mar/17/google-ministers-quiz-placement-ads-extremist-content-youtube.

233 **"We have a responsibility":** "Expanded Safeguards for Advertisers," Google blog, March 21, 2017, https://blog.google/technology/ads/expanded-safeguards-for-advertisers.

234 **surveyed dozens of creators:** Internet Creators Guild (@Internet CreatorsGuild), "Adpocalypse Survey Data Collection," YouTube, April 29, 2017, www.youtube.com/watch?v=KhcXnIpweQg.

234 **Within a couple months they rebounded slightly:** David Pakman Show (@thedavidpakmanshow), "Yes, YouTube Ad Boycott STILL Crushing David Pakman Show," YouTube, April 24, 2017, www.youtube.com/watch?v=22Enxv1m8rU.

235 **investigation on children's content:** Sapna Maheshwari, "On YouTube Kids, Startling Videos Slip Past Filters," *New York Times*, November 4, 2017, www.nytimes.com/2017/11/04/business/media/youtube-kids-paw-patrol.html.

235 **writer and artist James Bridle:** James Bridle, "Something Is Wrong on the Internet," Medium, November 6, 2017, medium.com/@jamesbridle/something-is-wrong-on-the-internet-c39c471271d2.

235 **bizarre and downright abusive:** "Bad Baby with Tantrum and Crying for Lollipops Little Babies Learn Colors Finger Family Sond ..," YouTube screen capture, 1:21, *New York Times*, April 11, 2017, https://static01.nyt.com/images/2017/11/04/business/05YOUTUBEKIDS-4/05YOUTUBEKIDS-4-jumbo.jpg.

236 **"I don't even have kids and right now":** Bridle, "Something Is Wrong on the Internet."

236 **"This, I think, is my point":** Bridle, "Something Is Wrong on the Internet."

237 **Susan Wojcicki even considered:** Mark Bergen, *Like, Comment, Subscribe* (New York: Viking, 2022), 313.

237 **"We've talked about the adpocalypse":** Philip DeFranco (@PhilipDeFranco), "Why the Adpocalypse Is Worse than Ever and the NYT under Fire for 'Normalizing Hate,'" YouTube, November 27, 2017, www.youtube.com/watch?v=FgPjX5hDuPo.

237 **"I make one video about my eating disorder":** Daniel Nodar (@epDannyEdge), "so i make one video about my eating disorder and my entire channel is demonetized forever but logan paul can

show a dead body and make fun of suicide and #1 on trending 😨," Twitter, January 1, 2018, twitter.com/epDannyEdge/status/948043689577807872.

239 **"It was the first time we realized":** Bergen, *Like, Comment, Subscribe*, 323.

240 **A new phrase began making the rounds:** Bergen, *Like, Comment, Subscribe*, 324.

240 **To try to suppress extremist content:** Bergen, *Like, Comment, Subscribe*, x.

240 **Mid-level creators followed:** Taylor Lorenz, "YouTubers Beg Fans: Leave Videos on in the Background," *Daily Beast*, January 18, 2018, www.thedailybeast.com/youtubers-beg-fans-leave-videos-on-in-the-background.

241 **"I noticed these kids":** Taylor Lorenz, "Who's Getting Rich off All These Loud Teen YouTube Stars? This Guy," *Daily Beast*, December 14, 2017, www.thedailybeast.com/whos-getting-rich-off-all-these-loud-teen-youtube-stars-this-guy.

242 **investigation in 2018 found:** Chris Stokel-Walker, "YouTube Has Turned into a Merch-Plugging Factory," *New York Magazine*, April 20, 2018, nymag.com/intelligencer/2018/04/jake-paul-and-logan-paul-are-youtube-merch-monsters.html.

242 **"YouTube is turning into a giant":** Stokel-Walker, "YouTube Has Turned into a Merch-Plugging Factory."

Chapter 18: Breakdown and Burnout

245 **"The thing about prank culture":** Taylor Lorenz, "The Teen Taking Back Practical Jokes from YouTube's Bros," *Daily Beast*, February 1, 2018, www.thedailybeast.com/the-teen-taking-back-practical-jokes-from-youtubes-bros.

246 **NPR highlighted Mills's work:** Laura Sydell, "The Relentless Pace of Satisfying Fans Is Burning out Some YouTube Stars," NPR, August 13, 2018, www.npr.org/2018/08/13/633997148/the-relentless-pace-of-satisfying-fans-is-burning-out-some-youtube-stars.

// NOTES //

247 **Daniel Keem, a YouTube institution:** Taylor Lorenz, "How DramaAlert Became the TMZ of YouTube," *Daily Beast*, January 18, 2018, www.thedailybeast.com/how-drama-alert-became-the-tmz-of-youtube.

247 **"Over the past five years":** Rebecca Jennings, "The Gossip Accounts Telling You Which TikTok Star Is Dating Which YouTuber," *Vox*, February 12, 2020, https://www.vox.com/the-goods/2020/2/12/21127014/famous-birthdays-charli-damelio-chase-hudson.

248 **"I realized there was a big gap":** Taylor Lorenz, "Custom Photo Filters Are the New Instagram Gold Mine," *Atlantic*, November 13, 2018, www.theatlantic.com/technology/archive/2018/11/influencers-are-now-monetizing-custom-photo-filters/575686/.

248 **Everything came to a head:** Lindsay Dodgson, "Why the Beauty Community on YouTube Is One of the Most Turbulent and Drama-Filled Places on the Internet," *Insider*, October 2, 2019, www.insider.com/why-beauty-youtube-is-full-of-drama-and-scandals-2019-5.

248 **"The image and its caption":** Lindsay Dodgson, "How YouTube's Beauty Community Fell Apart with an Explosive Feud Called 'Dramageddon,'" *Insider*, August 28, 2021, https://www.insider.com/dramageddon-youtube-jeffree-star-manny-mua-laura-lee-gabriel-zamora-2021-8.

248 **"YouTube channels that cover drama":** Dodgson, "How YouTube's Beauty Community Fell Apart."

250 **Jake Paul eventually confronted Kołodziejzyk:** Jake Paul (@jakepaul), "confronting internet bully cody ko..." YouTube, May 18, 2019, https://www.youtube.com/watch?v=xf7vX3D8_ME; Morgan Sung, "Jake Paul's Attempt at Calling out 'Cyberbully' Cody Ko Backfired Beautifully," *Mashable*, May 20, 2019, mashable.com/article/jake-paul-cody-ko-cyberbully.

251 **"If you're not actively creating":** Sydell, "The Relentless Pace."

251 **"This is all I ever wanted":** Elle Mills (@ElleOfTheMills),

"Burnt out at 19," YouTube, May 18, 2018, www.youtube.com/watch?v=WKKwgq9LRgA.

252 **"I've often discussed the pressures"**: Casey Neistat (@casey), "The Pressure of being a YouTuber," YouTube, May 28, 2018, www.youtube.com/watch?v=G38ixvYVNyM.

252 **"Constant changes to the platform's algorithm"**: Julia Alexander, "YouTube's Top Creators Are Burning Out and Breaking Down En Masse," *Polygon*, June 1, 2018, https://www.polygon.com/2018/6/1/17413542/burnout-mental-health-awareness-youtube-elle-mills-el-rubius-bobby-burns-pewdiepie.

253 **over four hundred high-profile influencers**: @BaddieLambily, "Opps! Kendall better delete this.... #fyrefestival," Twitter, April 28, 2017, https://twitter.com/BaddieLambily/Status/857866995936751616.

253 **(In Kendall Jenner's case, $250,000)**: Lizzie Plaugic, "Fyre Fest Reportedly Paid Kendall Jenner $250K for a Single Instagram Post," *Verge*, May 4, 2017, www.theverge.com/2017/5/4/15547734/fyre-fest-kendall-jenner-instagram-sponsored-paid.

253 **"It's not the same as it was"**: Taylor Lorenz, "Influencers Are Abandoning the Instagram Look," *Atlantic*, April 23, 2019, www.theatlantic.com/technology/archive/2019/04/influencers-are-abandoning-instagram-look/587803/.

254 **The Happy Place billed itself**: Laura Studarus, "Meet Happy Place, Los Angeles' Newest Instagram-Friendly Experience," *Forbes*, December 1, 2017, www.forbes.com/sites/laurastudarus/2017/12/01/meet-happy-place-los-angeles-newest-instagram-friendly-experience/?sh=54f1d804307b.

254 **But when it arrived in Boston**: Murray Whyte, "'Happy Place' Comes to Boston, and It's Hell," *Boston Globe*, April 3, 2019, www.bostonglobe.com/arts/art/2019/04/03/happy-place-comes-boston-and-hell/GLJ2mdvgp0P9cwr1tcwk5H/story.html.

254 **"Instagram vs. reality" photos**: Olivia Wheeler, "Millie Mackintosh Flaunts Her Toned Figure in Instagram vs Reality Pic,"

Daily Mail, April 16, 2019, www.dailymail.co.uk/tvshowbiz/article-6928577/Millie-Mackintosh-showcases-washboard-abs-peachy-posterior-Instagram-vs-reality-post.html.

254 **At Beautycon, a beauty festival:** Cheryl Wischhover, "The Biggest YouTube Beauty Secret Has Nothing to Do with Makeup," Racked, June 2, 2016, www.racked.com/2016/6/2/11828904/beauty-vlogger-youtube-ring-lights/; Todd Perry, "Fitness Blogger Forever Exposed the Difference between Real Life and Instagram in Photos," *Good*, December 11, 2020, https://www.good.is/articles/ig-versus-reality-exposed/; Francesca Gariano, "Nutritionist Makes a Statement with 'Instagram vs. Reality' Pics," TODAY.com, January 20, 2019, https://www.today.com/style/nutritionist-shares-powerful-instagram-vs-reality-pics-promote-self-acceptance-t147119; Elizabeth Holmes, "What I Learned at Beautycon, Where 'Everyone Wants to Be Extra,'" *New York Times*, July 29, 2018, https://www.nytimes.com/2018/07/29/insider/beautycon-beauty-trade-show-kylie-jenner.html/.

254 **Influencers also started actively speaking out:** Eve Peyser, "The Instagram Face-Lift," *New York Times*, April 18, 2019, https://www.nytimes.com/2019/04/18/opinion/instagram-celebrity-plastic-surgery.html/; Roséline Lohr, "Social Media Burnout & the End of the Online Influencer Phenomenon," TIG, July 26, 2018, www.thisisglamorous.com/2018/07/social-media-burnout-the-end-of-the-online-influencer-phenomenon.html/.

254 **many teens began going out of their way:** Joshua Bote, "What's the Deal with Huji Cam, This Year's Trendiest Photo App?" *New York Magazine*, July 11, 2018, nymag.com/intelligencer/2018/07/what-is-huji-cam-this-years-hottest-photo-app.html.

254 **"For my generation":** Taylor Lorenz, "The Instagram Aesthetic Is Over," *Atlantic*, April 23, 2019, https://www.theatlantic.com/technology/archive/2019/04/influencers-are-abandoning-instagram-look/587803.

255 **Anything that felt staged:** Leah White, "It's Time Someone Said

It: 'Candid' Influencers Are Just as Fake as the Old Ones," *Fashion Journal*, April 29, 2019, https://fashionjournal.com.au/life/candid-influencers-just-fake-old-ones/.

255 **a vehicle to vent in the captions:** Ruth La Ferla, "The Captionfluencers," *New York Times*, March 27, 2019, https://www.nytimes.com/2019/03/27/style/instagram-long-captions.html/; Taylor Lorenz, "How Comments Became the Best Part of Instagram," *Atlantic*, January 4, 2019, www.theatlantic.com/technology/archive/2019/01/how-comments-became-best-part-instagram/579415/.

256 **When she launched a podcast:** Melody Chiu, "YouTube Star Emma Chamberlain's Podcast Hits No. 1 in 50 Countries — All about the Latest Episode," *People*, April 18, 2019, people.com/celebrity/youtube-star-emma-chamberlain-new-podcast-number-one-50-countries/.

256 **She vlogged about:** Emma Chamberlain (@emmachamberlain), "Taking My Driving Test...," YouTube, July 15, 2017, www.youtube.com/watch?v=Nqv_vKMjiUQ.

256 **"I made that video going":** Emma Chamberlain (@emmachamberlain), "We All Owe the Dollar Store an Apology," YouTube, July 27, 2017, www.youtube.com/watch?v=Y5-6f1T9qsc/; Lauren McCarthy, "Creating Emma Chamberlain, the Most Interesting Girl on YouTube," *W*, June 10, 2019, https://www.wmagazine.com/story/emma-chamberlain-youtube-interview.

258 **"I've cried multiple times":** McCarthy, "Creating Emma Chamberlain."

259 **"I Was A Relatable Youtuber":** Don's Life (@donslife6455), "I Was A Relatable Youtuber For A Day! *cough* Emma Chamberlain," YouTube, December 13, 2018, www.youtube.com/watch?v=eSGEWXIzVzo&t=13s.

259 **"I feel like that [Emma Chamberlain] aesthetic":** Taylor Lorenz, "Emma Chamberlain Is the Most Important YouTuber Today," *Atlantic*, July 3, 2019, www.theatlantic.com/technol

ogy/archive/2019/07/emma-chamberlain-and-rise-relatable-influencer/593230/.

Chapter 19: TikTok Dominates

264 **It was too full of hate now:** Pippa Raga, "While Lisa and Lena Are No Longer on TikTok, Their New YouTube Channel Is a Must-Watch," Distractify, September 9, 2019, www.distractify.com/p/why-did-lisa-lena-delete-tiktok-account.

266 **"With every new TikTok star who dances":** Rebecca Jennings, "The Year TikTok Became Essential," *Vox*, December 8, 2020, www.vox.com/the-goods/2020/12/8/22160034/tiktok-top-100-bella-poarch.

267 **to launch the Hype House:** Hype House (@thehypehousela), "'Merry Christmas from the Hype House Photos by @Bryant,'" Instagram, December 25, 2019, www.instagram.com/p/B6gMaDaldmf.

267 **members of a YouTube collab channel:** Our2ndLife (@Our2ndLife), "O2L HOUSE TOUR IN 60 SECONDS," YouTube, February 13, 2014, www.youtube.com/watch?v=2aVoaohvQK8.

267 **YouTuber mansions dotted L.A.:** Danni Holland, "David Dobrik House: Exclusive Photos of The Vlog Squad House!" Velvet Ropes, April 19, 2021, https://www.velvetropes.com/backstage/david-dobrik-house; Taylor Lorenz, "Meet the Teens and Parents Who Spend Hours Standing in the Hot Sun Outside Jake Paul's House," Mic, July 31, 2017, www.mic.com/articles/183081/meet-the-teens-and-parents-who-spend-hours-standing-in-the-hot-sun-outside-jake-pauls-house/; "Clout House – Here's Who Lives in the Million Dollar Mansion," Vlogfund, May 13, 2018, https://www.vlogfund.com/en/blog/clout-house/; Alex Williams, "How Jake Paul Set the Internet Ablaze," *New York Times*, September 8, 2017, https://www.nytimes.com/2017/09/08/fashion/jake-paul-team-10-youtube.html/.

269 **the "Sway boys":** Hanna Lustig, "When 2 Famous TikTok House Members Were Arrested on Drug Charges, They Were

Road-Tripping across the US in a Controversial Mid-Pandemic Adventure," *Insider*, May 27, 2020, www.insider.com/bryce-hall-arrest-drug-sway-house-jaden-hossler-charges-2020-5.

269 **a collection of internet-savvy gamers:** Taylor Lorenz, "Can FaZe Clan Build a Billion-Dollar Business?" *New York Times*, November 15, 2019, https://www.nytimes.com/2019/11/15/style/faze-clan-house.html; @FaZe Clan, "Revealing the New $30,000,000 FaZe House," YouTube, March 16, 2020, www.youtube.com/watch?v=-PZQqZ6n25k.

269 **Clubhouse took the incubator model:** Taylor Lorenz, Peter Eavis, and Matt Phillips, "TikTok Mansions Are Publicly Traded Now," *New York Times*, November 20, 2020, www.nytimes.com/2020/11/20/style/clubhouse-tiktok-tongji-west-of-hudson.html.

271 **"we talked. no fighting. it's settled":** Jaden Hossler (@jxdn), "we talked. no fighting. it's settled. i was heated asf but now i'm calm bc talking can resolve everything. it's over," Twitter, July 7, 2020, twitter.com/jxdn/status/1280371053793951745.

271 **"This is a TV series waiting to be made":** Hemanth Kumar (@crhemanth), "This is a Tv series waiting to be made. Who's calling dibs on this one?," Twitter, July 8, 2020, twitter.com/crhemanth/status/1280879916934787074.

271 **Wheelhouse, a production studio:** "Wheelhouse Media | Video Production Studio - Charlotte NC," Wheelhouse Media, www.wheelhousemedia.tv/.

271 **The D'Amelio family landed:** Alexandra Jacobs, "The D'Amelios Are Coming for All of Your Screens," *New York Times*, August 28, 2021, https://www.nytimes.com/2021/08/28/style/charli-dixie-damelio-hulu-show.html.

Chapter 20: Unlocked

273 **ER nurses, doctors, and health workers:** Shira Ovide, "A TikTok Doctor Talks Vaccines," *New York Times*, December 14, 2020, www.nytimes.com/2020/12/14/technology/a-tiktok-doctor-talks-vaccines.html.

274 **Sick people chronicled:** Sarah Wildman, "My Daughter, TikTok Warrior," *New York Times*, December 29, 2020, www.nytimes.com/2020/12/29/opinion/sunday/cancer-tiktok.html.

274 **popularized viral food trends:** Naomi Tomky, "Pancake 'Cereal' Is Basically Homemade Cookie Crisp for Teens (and Parents)," Kitchn, May 7, 2020, https://www.thekitchn.com/pancake-cereal-tiktok-23035559/; Jesse Szewczyk, "I Tried the 'Whipped Coffee' Trend That's Taking Over the Internet. Here's How It Went," Kitchn, March 20, 2020, https://www.thekitchn.com/whipped-coffee-trend-review-23017225; Naomi Tomky, "This Wildly Popular Recipe Turns Carrots into 'Bacon' in Just 10 Minutes," Kitchn, April 23, 2020, https://www.thekitchn.com/vegan-carrot-bacon-tiktok-23030278/; Trilby Beresford, "Chef, TikTok Star Eitan Bernath Signs With WME (Exclusive)," *Hollywood Reporter*, May 4, 2020, https://www.hollywoodreporter.com/business/business-news/tiktok-star-youtuber-eitan-bernath-signs-wme-1292864/; Rachel E. Greenspan, "How a Teenage Chef Created a Social Media Empire with Millions of Views, from Quarantine Cooking to an Appearance on 'Chopped,'" *Insider*, March 17, 2020, https://www.insider.com/eitan-bernath-chef-food-influencer-tiktok-cooking-videos-2020-4.

274 **When schools ceased in-person education:** Amelia Nierenberg and Adam Pasick, "Streaming Kindergarten on TikTok," *New York Times*, September 18, 2020, https://www.nytimes.com/2020/09/18/us/remote-learning-tiktok.html.

274 **elaborate fake story lines:** Caroline Haskins, "Why Teens Love TikTok," *Vice*, 23 July 2019, www.vice.com/en/article/bj9qq5/this-meme-explains-why-tiktok-isnt-like-any-other-social-media.

274 **teenagers began cosplaying multinational corporations:** Taylor Lorenz, "Step Chickens and the Rise of TikTok 'Cults,'" *New York Times*, May 26, 2020, www.nytimes.com/2020/05/26/style/step-chickens-tiktok-cult-wars.html.

274 **Book recommendation videos:** Elizabeth A. Harris, "How Cry-

ing on TikTok Sells Books," *New York Times*, March 20, 2021, www.nytimes.com/2021/03/20/books/booktok-tiktok-video.html?searchResultPosition=2.

275 **OnlyFans was founded:** Shanti Das, "Meet the King of Homemade Porn—A Banker's Son Making Millions," *Sunday Times* (UK), July 26, 2020, https://www.thetimes.co.uk/article/meet-the-king-of-homemade-porn-a-bankers-son-making-millions-z9vhq9c9s.

275 **earned a collective $4 billion:** Kaya Yurieff, "OnlyFans' Sustained Pandemic Boom; Twitter Finally Tests Edit Button," Information Archive, September 1, 2022, archive.is/JXdRk#selection-609.0-609.68.

275 **It would have been naïve:** Charlotte Shane, "OnlyFans Isn't Just Porn," *New York Times*, May 18, 2021, www.nytimes.com/2021/05/18/magazine/onlyfans-porn.html.

275 **earned $3 billion since the launch:** K. C. Ifeanyi, "The NSFW Future of OnlyFans, Where Celebs, Influencers, and Sex Workers Post Side by Side," *Fast Company*, March 26, 2021, www.fastcompany.com/90611207/the-nsfw-future-of-onlyfans-where-celebs-influencers-and-sex-workers-post-side-by-side.

276 **Over 30,000 creators joined Patreon:** Sarah Perez, "Over 30K Creators Joined Patreon This Month, as COVID-19 Outbreak Spreads," *TechCrunch*, March 26, 2020, techcrunch.com/2020/03/26/over-30k-creators-joined-patreon-this-month-as-covid-19-outbreak-spreads/.

276 **raised a fresh round of funding:** Maria Armental, "Patreon Tops $1 Billion Valuation as Pandemic Brings a Surge in Creators to Platform," *Wall Street Journal*, September 1, 2020, archive.is/retji#selection-229.1-232.0.

276 **venture-backed "creator economy" start-ups:** Taylor Lorenz, "Everything on Social Media Is for Sale," *Atlantic*, November 27, 2018, https://www.theatlantic.com/technology/archive/2018/11/young-artists-and-producers-embrace-micro-monetizing/576682/.

276 **Teenagers with large TikTok audiences:** Anna Cafolla, "How Young People Are Using TikTok to Get Political This General Election," *Dazed*, December 12, 2019, https://www.dazeddigital.com/politics/article/47105/1/general-election-tiktok-memes-boris-johnson-jeremy-corbyn-tory-labour.

276 **getting more political:** Rebecca Jennings, "TikTok Never Wanted to Be Political. Too Late," *Vox*, January 22, 2020, www.vox.com/the-goods/2020/1/22/21069469/tiktok-memes-funny-ww3-politics-impeachment-fires.

277 **raised awareness about Mike Pence's track record:** Joseph Longo, "Welcome to Mike Pence's Gay Teen Summer Camp," *MEL Magazine*, January 6, 2020, melmagazine.com/en-us/story/camp-pence-tiktok-memes-lgbtq-conversion-therapy.

277 **The nickname "Mayo Pete":** Joseph Longo, "Teens on TikTok Are Roasting the Hell out of 'Mayo Pete,'" *MEL Magazine*, November 19, 2019, melmagazine.com/en-us/story/mayo-pete-memes-buttigieg-tiktok-teens.

277 **the threat of nuclear war with Iran:** Charlie Beall (@charlie91bea), "This TikTok has been really keeping me going lately. An interpretative dance of @SpeakerPelosi announcing impeachment inquiries," Twitter, October 19, 2019, twitter.com/charlie91bea/status/1185622860611674112; John Herrman, "Welcome to TrumpTok, a Safe Space from Safe Spaces," *New York Times*, May 13, 2019, www.nytimes.com/2019/05/13/style/trump-tiktok.html.

278 **began covering BLM protests:** Kellen Browning, "Where Black Lives Matter Protesters Stream Live Every Day: Twitch," *New York Times*, June 19, 2020, www.nytimes.com/2020/06/18/technology/protesters-live-stream-twitch.html.

278 **a BLM protest in Minnesota:** Kareem Rahma (@Kareemrahma), TikTok profile, www.tiktok.com/@kareemrahma.

278 **The stunt garnered mainstream attention:** Taylor Lorenz, Kellen Browning, and Sheera Frenkel, "TikTok Teens and K-Pop Stans Say They Sank Trump Rally," *New York Times*, June 21,

2020, www.nytimes.com/2020/06/21/style/tiktok-trump-rally-tulsa.html.

278 **the platform finally cracked down:** Bobby Allyn, "TikTok Tightens Crackdown on QAnon, Will Ban Accounts That Promote Disinformation," NPR, October 18, 2020, www.npr.org/2020/10/18/925144034/tiktok-tightens-crackdown-on-qanon-will-ban-accounts-that-promote-disinformation.

278 **However, other conspiracies, such as:** E. J. Dickson, "A Wayfair Child-Trafficking Conspiracy Theory Is Flourishing on TikTok, despite It Being Completely False," *Rolling Stone*, July 14, 2020, www.rollingstone.com/culture/culture-news/wayfair-child-trafficking-conspiracy-theory-tiktok-1028622/.

278 **When states were called on Election Night:** Taylor Lorenz, "Election Night on TikTok: Anxiety, Analysis and Wishful Thinking," *New York Times*, November 4, 2020, www.nytimes.com/2020/11/04/style/tiktok-election-night.html.

278 **the administration was holding press briefings:** Taylor Lorenz, "The White House Is Briefing TikTok Stars about the War in Ukraine," *Washington Post*, March 11, 2022, www.washingtonpost.com/technology/2022/03/11/tik-tok-ukraine-white-house/.

279 **"That's what drives this":** Cassie Miller and Rachel Rivas, "The Year in Hate & Extremism Report 2021," Southern Poverty Law Center, March 9, 2022, www.splcenter.org/20220309/year-hate-extremism-report-2021.

279 **stars like Lizzo:** Jules (@bangchannies), "CHAN AND FELIX TAKING THE RENEGADE DANCE SO SERIOUSLY SENDS," Twitter, February 7, 2020, twitter.com/bangchannies/status/1225950060569075712.

279 **She listened to the beats in the song:** Dance Tutorials Live (@DanceTutorialsLive), "ARM WAVE TUTORIAL | How to Dance to Dubstep: WAVING» Beginner Hip Hop Moves W/ @MattSteffanina," YouTube, March 23, 2013, www.youtube.com/watch?v=6CPtOe3GVwk/; Eli Unique (@eliuniquee), "HOW to DO the WOAH DANCE & PLENTY of WAYS !!!"

YouTube, July 7, 2018, www.youtube.com/watch?v=ZPNfN63WgXw.

280 **Creators of popular dances:** "Backpack Kid Russell Horning, Creator of the Floss Dance, Becomes the Latest to Sue Fortnite," ABC News (Australia), December 18, 2018, https://www.abc.net.au/news/2018-12-19/floss-dance-creator-backpack-kids-sues-fortnite/10633962; Allie Yang, "Shiggy on How the 'in My Feelings Challenge' Changed His Life," ABC News, December 21, 2018, abcnews.go.com/Entertainment/shiggy-feelings-challenge-changed-life/story?id=59782945.

280 **@TikTok Room picked up the controversy:** Taylor Lorenz, "The Original Renegade," *New York Times*, February 13, 2020, www.nytimes.com/2020/02/13/style/the-original-renegade.html/; D1 Nayah (@thereald1.nayah), Instagram profile, www.instagram.com/thereald1.nayah/.

280 **Despite creating and driving:** Ashley Carman, "Black Influencers Are Underpaid, and a New Instagram Account Is Proving It," *Verge*, July 14, 2020, www.theverge.com/21324116/instagram-influencer-pay-gap-account-expose.

280 **Accounts like @InfluencerPayGap called attention:** Tiffany Trotter, "New Instagram Account Exposes Pay Disparities among Black Influencers," *Black Enterprise*, August 18, 2020, www.blackenterprise.com/new-instagram-account-exposes-pay-disparities-among-black-influencers/.

280 **Black creators were featured less frequently:** Imogen Learmouth, "The Dote Scandal and How It Reflects YouTube's Racism Problem," Thred Website, May 17, 2019, thred.com/culture/the-dote-scandal-and-how-it-reflects-youtubes-racism-problem/.

280 **disproportionately overlooked:** Amanda Silberling, "Atlanta-Based Black Influencer Collective Swapped Collab House for Studio," *TechCrunch*, August 5, 2022, techcrunch.com/2022/08/05/collab-crew-black-influencer-collective-studio-atlanta/.

280 **Keith Dorsey, a talent manager:** Lorenz, "The Original Renegade."

281 **They were also regularly featured:** Jenna Wortham, "Instagram's TMZ," *New York Times*, April 14, 2015, www.nytimes.com/2015/04/19/magazine/instagrams-tmz.html.

281 **"Racism even plays into the algorithm":** Taylor Lorenz, "The New Influencer Capital of America," *New York Times*, December 11, 2020, www.nytimes.com/2020/12/11/style/atlanta-black-tiktok-creators.html/; Nicky Campbell, "CFDA," Cfda.com, July 27, 2020, cfda.com/news/chrissy-rutherford-and-danielle-prescod-launch-consulting-agency-2bg.

281 **"We are being exploited":** Taylor Lorenz and Laura Zornosa, "Are Black Creators Really on 'Strike' from TikTok?" *New York Times*, June 25, 2021, www.nytimes.com/2021/06/25/style/black-tiktok-strike.html.

281 **By mid-2023, the industry:** Brianna Holt, "Black Content Creators Receive Less Money than Their White Counterparts. They Are Relying on the Strength of the Creator Community to Lessen the Pay Gap," *Insider*, January 8, 2023, https://www.insider.com/how-black-content-creators-trying-to-lessen-influencer-pay-gap-2023-1/; Daysia Tolentino, "Black Creators Say They 'Have to Be Perfect' to Get Promotional Products from Brands. They Want That to Change," NBC News, December 27, 2022, www.nbcnews.com/news/nbcblk/black-creators-call-out-inequity-influencer-gifting-rcna61923.

282 **"A lot of brands have complained":** Lorenz, "The New Influencer Capital of America."

Chapter 21: The Scramble and the Sprawl

284 **invested over $2 billion:** "TI Creator Economy Database," *Information*, https://www.theinformation.com/creator-economy-database.

284 **those numbers expanded:** Kaya Yurieff, "What We Learned from TI's Creator Economy Database," Archive.ph, June 28, 2021, archive.ph/Gg4Eh.

284 **Clubhouse raised $200 million:** Alex Heath, "Briefing: Club-

house Confirms New Funding Round Led by Andreessen Horowitz," *Information*, April 18, 2021, www.theinformation.com/briefings/ca4cc3.

284 **Jake Paul announced:** Edward Ongweso, "Jake Paul Is Turning His Massive Audience into Fodder for His New VC Fund," *Vice*, March 31, 2021, www.vice.com/en/article/n7v87g/jake-paul-is-turning-his-massive-audience-into-fodder-for-his-new-vc-fund.

284 **Animal Capital:** Ongweso, "Jake Paul Is Turning His Massive Audience into Fodder."

285 **an investigation by the journalist Kat Tenbarge:** Kat Tenbarge, "A Woman Featured on YouTube Star David Dobrik's Channel Says She Was Raped by a Vlog Squad Member in 2018 the Night They Filmed a Video about Group Sex," *Business Insider*, March 16, 2021, www.businessinsider.com/vlog-squad-durte-dom-rape-allegation-david-dobrik-zeglaitis-video-2021-3.

285 **Before it stopped paying its bills:** Taylor Lorenz, "A TikTok Rival Promised Millions to Black Creators. Now Some Are Deep in Debt," *Washington Post*, August 3, 2022, www.washingtonpost.com/technology/2022/08/01/triller-app-black-creators-pay/.

285 **a program to split advertising revenue:** Garett Sloane, "Meta Puts New Ads in Facebook Reels and Will Share Revenue with Creators," *Ad Age*, October 4, 2022, adage.com/article/digital-marketing-ad-tech-news/meta-puts-new-ads-facebook-reels-and-will-share-revenue-creators/2438186.

285 **a program called Pulse:** Aisha Mailk, "YouTube Rolls out New Partner Program Terms as Shorts Revenue Sharing Begins on February 1," *TechCrunch*, January 9, 2023, techcrunch.com/2023/01/09/youtube-new-partner-program-terms-shorts-revenue-sharing-february-1/.

286 **despite a slowdown in funding:** Mahira Dayal, "The Creator Economy's Next Chapter: Living with Less," *Information*, January 3, 2023, https://www.theinformation.com/articles/the-creator-economys-next-chapter-living-with-less.

286 **Goldman Sachs predicts:** "The Creator Economy Could Approach Half-a-Trillion Dollars by 2027," Goldman Sachs, April 19, 2023, https://www.goldmansachs.com/insights/pages/the-creator-economy-could-approach-half-a-trillion-dollars-by-2027.html.

286 **"A lot of people still see YouTubers":** Chloe Sorvino, "Could MrBeast Be The First YouTuber Billionaire?" *Forbes*, November 30, 2022, https://www.forbes.com/sites/chloesorvino/2022/11/30/could-mrbeast-be-the-first-youtuber-billionaire.

// INDEX //

Abascal, Lina, 16, 94, 115
ABC, 14
Abuhamdeh, Tayser (Mr. Cashier), 161–162
Abu-Taleb, Yousef, 64, 65
Ackerman, Spencer, 175
Adesanya, Abby, 259
Adler, Ben, 175
Adpocalypse, 231–235, 237, 240, 243
ad-revenue sharing, 196, 285–286
adult entertainment, 274–276
advertising, 15, 22–25, 34, 36, 67, 70–78, 80–82, 89–90, 97, 120, 129–135, 139, 166–168, 182, 186, 195, 201, 215–219, 229–243
    see also monetization
affiliate marketing programs, 25, 137–139, 221, 286
aggregators, 13, 120–122, 167
Ahanonu, Monica, 225
Akon, 140
Alfred's, 224
algorithm, 231–233, 236–237, 251, 265–266

Allison, Julia (Julia Allison Baugher), 41–42, 46–55, 78, 103, 111, 250
"All I Want for Christmas" (song), 242
Amazon, 11, 161, 192, 207, 286
"America, Fuck Yeah" (song), 116
American Prospect, 12, 13
AM New York, 42, 46
Anderson, Steve, 129
Anderson, Tom, 31
Andreessen Horowitz, 284
Animal Capital, 284
Annon, Kouvr, 268
Annoying Orange (channel), 88, 91
antisemitism, 229–230
Anys, Imane (Pokimane), 159
Aokigahara forest, video from, 237–240
AOL, 11, 102
Apple, 275
Arment, Marco, 48
Armstrong, Heather, 20, 22–23, 25–27
artificial intelligence (AI), 236

ASCII art, 11
AsToldByKenya, 258–259
Atendy, 147
AT&T, 163, 164, 233
Audience (firm), 182
"Auto-Tune the News" (web series), 78
Avery, Bree, 63–65
AwesomenessTV, 163
Ayala, Robby, 167

Baby Ariel, *see* Martin, Ariel
Bachelor, Andrew (King Bach), 176–178, 183–185, 192, 196
Bahner, Adam (Tay Zonday), 68–71, 75, 76, 83
Banana Republic, 25, 26
Bannon, Steve, 165
Baugher, Julia Allison, *see* Allison, Julia
TheBdonski, 79
Beam, Alex, 103
Bearman, Erika, 110–113
Beautycon, 254
Beck, Noah, 284
Beckett, Miles, 63, 64
"Bed Intruder Song" (song), 88
Bedrocket Media Ventures, 166
Bennett, Graham, 239
Bennett, Jessica, 117
@Bergdorfs (Twitter account), 110
Bernath, Eitan, 274
Beutler, Brian, 175
2BG Consulting, 281

B&H Photo Video, 88
bias, against Black creators, 279–282
Biden, Joe, 276, 278
Bieber, Justin, 130–131, 176, 177, 269
Big Tech, 1, 207, 285–286, 291
Bilton, Nick, 102, 104, 105
Bixler High, Private Eye (film), 264
Black creators, 258–259, 269, 278–282
Black Lives Matter (BLM) protests, 278
Blakeley, Richard, 49
Blast, 129
BlogAds, 15, 22
blogger gate, 16
BlogHer event, 24
blog roll, 21
blogs, 2–7, 11–17, 22–27, 46–51, 113, 135–139, 175
see also mommy bloggers; Tumblr
Blogsnark, 247
Blogspot, 19
Bloomberg, Michael, 49
Blutstein, Reese, 254–255
Boedigheimer, Dane, 88
#BookTok, 274
Bordelon, Bart, 168–169
Bosstick, Michael, 223
Box, Baxter, 136, 137
brands, 108–113, 132–133, 219–222, 270
see also sponsored content deals

// INDEX // 357

Braun, Scooter, 130
Bravo, 51, 52
Brence, Jon, 251
Brennan, Amanda, 116
Britton, Evan, 248
broetry, 200
Buca di Beppo, 171–172
bullying, 32, 249–252
Bundesen, Tabatha and Bryan, 94, 95
Burbn, 128
burnout, 251–252, 255
Burrows, Kevin, 118
Buttigieg, Pete "Mayo Pete," 277
BuzzFeed, 49, 120–123, 194, 200
ByteDance, 207, 263

Cabalona, Jeremy, 178–182, 185
Cabin Six, 277
cable television, 43
Camahort Page, Elisa, 24
Cam Girls, 161
Capps, Kriston, 175
Captiv8, 211
Caramanica, Jon, 198
Carbone, Lexie, 254
Carl's Jr., 80–82
Carpenter, Aaron, 168–170
Carrot Creative, 168
Cartoon Network, 88
cat videos, 71–73
CBS Films, 167
celebrity and celebrities, 16, 29, 38–55, 68–73, 95, 105, 108–113, 117–119, 123–125, 128–132, 139–141, 153, 159, 172, 183, 198, 200, 209, 253, 291
    see also internet stardom and stars
Cerny, Amanda, 177, 193
ceWEBrities, 50
Chamberlain, Emma, 255–259, 286
Charnas, Arielle, 220–223
Chasing Cameron (Netflix series), 171
"Cheerleader" (song), 205–206
Chen, Steve, 59
Chernin Group, 163, 164
"Chocolate Rain" (song), 69–70
Christian Dior, 224–225
Cicada, 153
Cicero, Nick, 217
Cîroc, 215
Cisco, 51
City, The (TV series), 5
Clash, 270
Clickbait (merchandise), 242
Clift, Simon, 51
Clout Gang, 211
Clubhouse (content house), 269, 271
Clubhouse app, 284
Cobrasnake, The (blog), 16
Coca-Cola, 167, 226
Cohen, Taylor, 253
collab channels, 76–80, 82–83, 87
Collab Crib, 281, 282
collab houses, 51, 79–80, 175–180, 267–272, 277, 281–282
Collective, The, 91
Collective Digital Studio, 91
comedy videos, 147–149, 184–185

Comments by Celebs, 247
compensation, 225–227, 275,
    285–287
competition, among video apps,
    151–154, 193
connector moms, 23–24
Connors, Catherine, 20, 26, 27
Consumer Electronics Show, 51
consumer goods, 132–133
content farming, 232
content houses, *see* collab houses
"Cool Cat" (video), 71
Cooperstein, Ezra, 81–82, 91
copyright infringement lawsuits, 61,
    67
Costolo, Dick, 147
Covid-19 pandemic, 55, 273–278,
    283, 286, 290
Creative Artists Agency (CAA), 65
Creator Clubs, YouTube, 208
creator economy (influencer
    economy), 53–54, 89–97,
    210–213, 273, 283–287, 290
creators, 85–97
    advertising for, 70, 72–73
    Black, 258–259, 269, 278–282
    code of conduct for, 239–240
    compensation for, 225–227, 275,
        285–287
    demonetization for, 229–231,
        233–235, 237–240
    empowering, 200
    entertainment industry and, 162,
        163
    Grumpy Cat meme, 94–97
    harassment and bullying of,
        250–252
    influence of, 191, 289
    on Instagram, 133–135, 140
    as liability, 231–235
    merchandise deals for, 240–243
    on mobile video apps, 154–156
    mommy blogger model for, 27
    at multichannel networks,
        89–93
    performance metrics for, 93–94
    platform relationships with, 29,
        70, 82–83, 87–89, 162–164,
        179–188, 197–200, 208–209,
        289–290
    products driven by, 223
    radicalization for, 164–165
    relationships between fans and,
        85–87
    schedule for, 243, 245–247,
        250–251, 258, 268
    social networks for, 39
    touring by, 168–173
    on Twitter, 110
    use of term, 83, 92–93
    venture capital firms of, 284–285
    VidCon event for, 85–87
    in YouTube Partner Program,
        67, 72, 74–75
    see also influencers
cringe content, 249–250, 266
cross-platform sharing, 280
crowning, 155–156

Cruikshank, Lucas (Fred Figglehorn), 88, 89, 91
#cutformagcon, 171
Cutshall, Matt, 168
Cyr, Kio, 269

DaddyOFive (channel), 230–231
Dailey, Dustin, 249
Daily Dot, 123
daily vlogs, 209–211
Dallas, Cameron, 170, 171, 173
D'Amelio, Charli, 265, 267, 271, 285, 286
D'Amelio, Dixie, 265, 267, 271
D'Amelio Show, The (Hulu series), 271
Danielle, Anthony, 134
Darth Vader (Twitter account), 105
Davis, Peter, 5
Dawson, Shane, 79, 87, 163, 195
Days, Dave, 79
Dear Media (podcast), 223
DeFranco, Philip, 79, 87, 234
DeGeneres, Ellen, 140–141, 148, 151
Delevingne, Cara, 141
Delmondo, 217
demonetization, 229–231, 233–235, 237–240
Denton, Nick, 47
des Jardins, Jory, 24
Deuxmoi, 247
DeWolfe, Chris, 31
Diet Prada, 247
DiFeo, Brian, 134
Digg.com, 69

Digital Brand Architects (DBA), 223
digital cameras, 30, 128
digital reputation, 290
digital shorts, 60–61
digital television, 36
DigiTour, 156, 169–173, 206, 208
disinformation (fake news), 231–232, 278
Disney Digital Network, 164
Dispo, 284–285
diversification, 199, 263–265, 276
Diversity University, 277
DKNY PR GIRL, 108–113
D1 Nayah, 280
Dobre brothers, 249–250
Dobrik, David, 210, 242, 284–285
Dodson, Antoine, 88
Dolce, Christine (ForBiddeN), 36
dollar store haul videos, 256–257
Donaldson, Jimmy (MrBeast), 223, 286–287
Don Bbw, 259
Donna Karan, 108–113
Donovan, Hannah, 188–189
Donovan, Joan, 165
Donovan, Lisa (LisaNOVA), 75, 77, 79, 80, 82, 90
"Don't Judge Challenge," 205–206
Donut Shop dance, 280
Dooce.com, 20, 22–23, 25
Dore, Garance, 16
Dorsey, Jack, 102–105, 128, 188, 189
Dorsey, Keith, 280–281
Douyin, 207

Dr. Pepper, 70
Dragun, Nikita, 248
Drake, 159
DramaAlert (channel), 247
drama channels, 247–250, 266, 270
Dramageddon, 248–249
DreamWorks Animation, 163
Dress, The, on Tumblr, 121–123
Dreyfuss, Emily, 165
Drudge, Matt, 13
Drudge Report, 13, 115–116
Drummond, Ree, 24
Dubsmash, 279, 280
Ducard, Malik, 91

Easterling, Addison, 265, 267
Edwards, Kate, 220
e-lebrities, 111–112
Elsagate, 235–237
endorsements, 133–134
Engel, Julia, 223
Enterprise, 233
entertainment industry, 39–40, 140–141, 162–163, 172, 183, 192, 264, 271–273
Erickson, Rawn, 79
Eswein, Liz, 127, 129–135
extremism, 164–165, 232–233, 240, 279

Facebook, 29, 32–39, 48–49, 55, 95, 101, 103, 113–114, 120–122, 131–132, 135, 140, 161, 188, 193–197, 207, 231–232, 285
Facebook Live, 193–197
Fairway, 134
fake sponsored content, 224–225
fame, *see* celebrity and celebrities
Famous Birthdays, 247
Fanjoy, 241–242
"Fanjoy to the World" (song), 242
fashion industry, 32–33, 108–113, 118–119, 124, 132–133, 136–139, 219–222
Fatemi, Falon, 87
FaZe Clan, 269
F8 conference, 194
Federal Trade Commission (FTC), 215–219, 225
Federated Media, 23, 26
feuds, 210–211, 249–250
filters, 127, 128, 129, 254
Fishman, Rob, 166, 167
Fitzpatrick, Alexandria (AlliCattt), 192
Fiverr, 230
Flaherty, Vanessa, 223
Flannery, Michele, 68
Flash animation, 11, 60
Flickr, 107
Flinders, Mesh, 63, 64
Flipagram, 207
Flophouse (blog house), 175
Floyd, George, 276, 278
Fohr and Fohr Card, 124, 217, 255
follower numbers, 139–140, 218, 281, 282
followers (term), 101, 106–107

"follow for follow" campaigns, 139–140
Foremski, Matthew, 65
Fortnite (video game), 159
For You page, TikTok, 265–266
founders, 31, 34, 38–40, 102–106, 146, 151, 179–180, 183, 200, 289
4chan, 69–70, 164
Fox, 44
Frazier, Darnella, 278
Freak Family, 235–236
Fred: The Movie (film), 88, 89
Frederator Studios, 92
Fred Figglehorn, *see* Cruikshank, Lucas
Freeman, Cassie, 222
free products, 23–24, 93, 136
    See also sponsored content deals
Friedberg, Brian, 165
friends, connecting with, 29, 103–104, 106, 140, 153
Friendster, 30–31, 33, 101, 140
Frier, Sarah, 129, 139
Frumin, Michael, 48
"Fuck Yeah" Tumblrs, 115–119
Fullscreen network, 91, 162–164, 171
Furlan, Brittany, 167
Fyre Festival, 253

Gamergate, 164–165
gaming, 37–38, 89, 91, 158–159, 164–165
Gandhi, Avi, 163
Gap, 168
Garner, Nate, 239
Garza, Alexandrea, 222
gatekeeping, 12, 15, 17, 23, 88–89, 274–275, 290
Gawker, 47, 51–52
General Electric, 166
Generation Z, 161, 205, 225, 254–255, 265–272, 277, 278
Georgetown University, 41
Get Off My Internets (GOMI), 36–37, 247
Gevinson, Tavi, 215
Gharaibeh, Kassem, 77, 79, 87, 163, 164
Gibsonlook, 222
GIFs, 146
Girls' Night In Tour, 171
Glass, Noah, 102
TheGlobe.com, 30
Goldman Sachs, 286
Gomez, Selena, 130, 207
Gonzalez, Philippe, 130
Good American, 216
Goodfried, Amanda, 65
Goodfried, Greg and Amanda, 63
Google, 33, 59, 67, 74, 76, 80, 81, 91, 161, 182, 207, 233, 275
Google AdSense, 15, 75
Google Preferred, 209
Google Video, 61–62, 66–67
Gore, Al, 6, 12–13
gossip blogs and channels, 36, 247–250
Gossip Girl (TV series), 109

#gpoy, 46
Graff, Garrett, 15–16
Grande, Ariana, 207
GrapeStory, 166
Gray, Loren, 264–265
Green, Hank, 85–87, 234
Green, John, 85–87
Green, Michael, 91
Green Acres (TV series), 44
Gregory Brothers, 78, 86, 88, 169
Greycroft, 89
Grier, Hayes, 171
Grier, Nash, 166, 170, 171, 173, 179
Griggs, Quinton, 269
Grimmie, Christina, 172–173
Grind, The (TV series), 44
group texting, 103
GRP Partners, 89
Gruger, William, 185, 188
Grumpy Cat, 94–97
Gutierrez, Manny (Manny MUA), 248

Hall, Bryce, 269
HallowMeme, 50
Hanes, 167
Hanna, Gabbie, 187, 192
Happy Place, 253–254
harassment, 26–27, 36–37, 171, 232, 236, 249–252
Harmon, Jalaiah, 279–281
Harris, Josh, 157
Hasan, Sheeraz, 42
hashtag feature, Twitter, 107
haul videos, 256–257

Hepburn, Ned, 117
h3h3productions (channel), 234
Hilton, Conrad, 42
Hilton, Paris, 42–46, 54, 80
Hilton, Perez, 16
Hobart House, 175
Hofmann, Alex, 153, 155, 206
Hofmann, Dom, 146, 147, 151
Holderness, Cates, 117, 121–123
homepage takeover, YouTube, 70
Hossler, Jaden, 269, 271
Houghton, Jeff, 213
House That Nobody Asked For, 269
HP, 167
Hudson, Chase, 267–268, 270
Huffington, Arianna, 166
Huji Cam, 254
Humans of New York (blog), 119
Hunter, Mark, 16
Hurley, Chad, 59, 67, 74
Hype House, 267–271
hype houses, political, 277

iCarly (TV series), 88
ICM, 271
@Igers (Instagram account), 130
Imgur, 94
India, 230
influence metrics, 93–94
influencers, 35, 38, 45, 50–51, 111, 123, 132, 135, 137–139, 165, 213, 219–227, 232, 249–250, 252–255, 284–287
see also creators

Instagram, 55, 96, 127–141, 166, 180
    advertising on, 139
    bloggers on, 135–139
    brands' use of, 132–133
    celebrities on, 128–132
    celebrity news on, 271–272
    compensation for creators on, 225–227
    creators' relationship with, 39, 133–135, 201–202
    credit for shared content on, 279, 280
    DigiTour for stars from, 172
    early marketing of, 38
    and Facebook, 207
    founding of, 127–128
    FTC crackdown on ads, 215–219
    gossip and drama channels on, 247
    influencer marketing on, 219–222
    mommy bloggers on, 25
    monetization on, 134–140, 168, 197
    Musical.ly and, 206–207
    and nightlife blogs, 16
    open networks on, 101
    resharing on, 178
    self-parody on, 252–255
    Snapchat and, 199
    sponsored content on, 119, 224–225
    transformation of internet by, 140–141
    Tumblr and, 124
    Vine and, 147, 182, 188, 192, 193, 211–231
    visual content on, 146
@instagram (account), 132
Instagramers.com, 130
Instagram husbands, 212–213
Instagram rapture, 139–140
Instagram Reels, 285
Instagram Video, 152
Instagram walls, 212
internet, 6–7, 11, 140–141, 265, 289, 291
Internet Creators Guild, 234
internet stardom and stars
    gossip and drama channels, 247–248
    Grumpy Cat, 94–96
    influence of, 192
    on Instagram, 140, 141
    managers for, 73
    MCNs treatment of, 163–164
    multi-platform, 161–162, 175–176, 188, 201, 252, 265, 267
    on MySpace, 35–37
    right-wing, 165
    on TikTok, 265–267, 270–271
    tours by, 168–173
    on Tumblr, 123–125
    on Twitch, 158–159
    at VidCon, 85–87
    video creators, 154–156
    on Vine, 147–150, 154, 165–168
    on YouTube, 92–93, 239–240
    see also celebrity and celebrities

Internet Week, 49, 121
interstitial ads, 201
"Into You" (song), 207
investors, 30, 89–90, 127, 172, 173, 206, 283–285
iPhone, 101, 127–128
Isba organization, 233
Issa, Caroline, 219
IT Girls (TV series), 51
"It's Everyday Bro" (song), 211
"It's Every Night, Sis" (song), 211
iTunes, 70, 88, 102

Jack and Jack, 172
January 6 insurrection, 291
Jarre, Jérôme, 147–149, 151, 154, 165–166, 168
Jenner, Kendall, 139, 253
Jenner, Kris, 45
Jenner, Kylie, 139, 216
Jezer-Morton, Kathryn, 26, 27
Jibaw, Anwar, 177
Jin, Li, 283
Johns, Marcus, 148–150, 166, 167, 178, 184–187
Johnson, Griffin, 269, 284
Johnson & Johnson, 233
Joshi, Palak, 224
Just Chatting category, Twitch, 277
Justin.tv, 157, 158

Kan, Justin, 157
Kantor, Andrew, 103
Karan, Donna, 112

Kardashian, Khloe, 216, 286
Kardashian, Kim, 43, 45–46, 216, 286
Kardashian, Kourtney, 216, 286
Karim, Jawed, 59, 60
Karp, David, 48, 115
K-Camp, 279
Keech, Daisy, 268, 269, 284
Keem, Daniel (Keemster), 247
Keeping Up with the Kardashians (TV series), 45
Keiko, Laurie, 130
Kennedy, Cory, 16
Keyboard Cat, 71–73, 163
Khaled, DJ, 198, 215
Kids portal, YouTube, 235–237
"KILLING BEST FRIEND PRANK" (video), 211
Kimojis, 216
King, Zach, 167, 265
Kirkpatrick, David, 114
Kjellberg, Felix (PewDiePie), 229–232, 234, 242
Klein, Ethan, 234, 235
Klein, Ezra, 175
Klein, Hila, 234
Klein, Matt, 255
Klein, Sharon, 44
Kloss, Karlie, 141
Klout score, 93–94
Know Your Meme, 49–50
Kolodziejzyk, Cody (Cody Ko), 249–250
Krieger, Mike, 127, 128
Krizelman, Todd, 30

Kroll, Colin, 146, 151, 179, 180
KSI, 242
Kumar, Hemanth, 271
Kutcher, Ashton, 43, 71, 108, 112, 133, 182, 198
Kylie Cosmetics, 216

Lachey, Nick, 43
Lachtman, Darren, 166, 167
Lashes, Ben, 71–73, 94–96, 163
Later Media, 254
"Lazy Sunday" (video), 60–61, 63
Leding family, 44
Lee, Laura, 248
Lewis, Kayla, 259
LGBTQIA+ TikTokkers, 277
liberal ideology, 164–165
Licht, Aliza, 108–113
lifecasting, 50
lifestyle bloggers, 138–139
LIKEtoKNOW.it (LTK), 138, 139
Lilhuddy, 268
LinkedIn, 200
Lisa and Lena, 264
LiveJournal, 102
Live.ly, 206
live-streaming apps, 157–158, 161–162, 206
Llanos, Ibai, 159
lo-fi content, 254–259
Lonelygirl15 series, 62–66, 75, 209
loop counts, 148, 180–181
Lord &Taylor, 215, 216
Lott, Trent, 14–15

"Lottery" (song), 279
Louis, Erick, 281
"Love" (song), 68–69

Machinima, 89, 91, 164
Magcon, 168–172
Maker Studios, 79–83, 89–91, 162–164, 229, 230
Mancuso, Rudy, 145–146, 149–151, 154, 167, 168, 177, 191, 192, 199
Marbles, Jenna, 163
marginalized identities, people with, 1, 24, 69, 277–279
Marshall, John, 12–15
Martin, Ariel (Baby Ariel), 154–156, 169, 263–265
Martin, Michael and Heather, 230–231
Mase (rapper), 140
Mashable, 95, 120, 123, 178–179
Maverick (merchandise), 241, 242
McBling style, 42–43
McBride, Shaun (Shonduras), 154, 198–202
McDonnell, Charlie, 86–87
McGlynn, Nick, 49
McKeen, Summer, 258
McMullan, Patrick, 4
McNeill, Caitlin, 121–123
media, legacy, 12–17, 19–21, 23, 26–27, 51–53, 67, 86, 88–89, 95, 96, 103, 108, 119, 123–125, 178, 201, 248, 249, 273, 287
Mediakix, 201
Meerkat, 158, 161

meet-and-greets, 171–173, 246–247
   see also DigiTour; Magcon
meet-ups, 49, 129, 149–150
Megalis, Nicholas, 148–150, 167
Melberger, Chris, 149
memes, 49–50, 68, 73, 94, 95, 119, 125, 198, 212, 277
mental health, 20, 25–27, 251–252, 264
merchandise deals, 240–243, 246
Merjos, Hanna Beth, 35
Messina, Chris, 107
Michael Kors, 139
microblogging, 102
micro-fame, 7, 47–48, 76
micro-influencers, 213, 222
Microsoft, 59, 72
Milan Fashion Week, 16
Millennial pink, 212
Millennials, 46, 110, 147–148, 212, 254
Mills, Elle, 245, 250–251
Minaj, Nicki, 207
Miracle on the Hudson, 107–108
misogyny, 51–53, 164–165, 250
Miss Advised (TV series), 52
Mobile Media Lab, 134, 135
mobile phones, 101, 102, 127–128
Mohney, Chris, 47
mommy bloggers, 19–27, 250
monetization
   and Adpocalypse, 234–235
   by Julia Allison, 50–54
   for Black creators, 259, 280–282
   for bloggers, 23–27, 135, 136–139
   in creator economy, 273
   diversification of, 240, 242
   on Facebook Live, 195–197
   on Instagram, 129–131, 134–140, 217
   of live-streaming, 161–162, 206
   for MCNs, 163–164
   on Snapchat, 199
   on social networks, 35–37
   on Tumblr, 118–119
   on Twitch, 159
   on Vine, 186–188
   on YouTube, 65–66, 71–76, 81, 196, 208–209
   see also specific types
Mongeau, Tana, 271
monopolization, 178, 184–185
Mortimer, Tinsley, 3–5
Mosallah, Arya, 245
Moss, Kate, 219
MrBeast, see Donaldson, Jimmy
@M3ssyM0nday (Instagram account), 272
MTV, 43–44, 93–94
multichannel networks (MCNs), 82–83, 89–93, 162–164, 166, 171–172, 208, 241
   see also specific networks
multi-platform internet stars, 161–162, 175–176, 188, 201, 252, 265, 267
museums, Instagram, 212

Musical.ly, 153–156, 158, 169, 172, 180, 188, 192, 194, 205–208, 217, 263–267
music blogs, 16
music challenges, 205–206
music industry, 206–207
music videos, 68–71, 76
My Damn Channel, 71, 89
MySpace, 29, 31–40, 48, 63–65, 70, 78, 120, 250
MySpace TV, 36

Napster, 61
NARS, 225
NBC Universal, 61
Neistat, Casey, 209–210, 251–252
network effects, 82
Neutrogena, 66
Newlyweds (TV series), 43
News Corp, 33, 37
News Feed, Facebook, 39, 113–114
newspapers, 16–17
@newyorkcity (Instagram account), 127, 132, 133
New York Fashion Week, 16, 124
Next New Networks, 48, 77–78, 82–83, 88, 89, 92, 208
Nguyen, MaiLinh, 256, 257
Niche (agency), 166–168
Nies, Eric, 43–44
nightlife blogs, 16
Night to ReMEMEber, A, 49–50
Nike, 134
Ninja (Twitch creator), 159
Non Society, 51
Nord, James, 123–124, 217–219, 225, 255
Nordstrom, 134, 219–223

Oakley, Peter (geriatric1927), 66
Oakley, Tyler, 87
Obama, Barack, 105, 113
Obama, Michelle, 183
O'Connor, Ciaran, 279
Odeo, 102
O'Farrell, Brad, 71, 72
offensive content, 184–185
O2L and O2L Mansion, 172, 267
Oliver, Jamie, 131
Olivia Palermo Beauty, 6
OMI, 205–206
Omnivision Entertainment, 71
O'Neal, Shaquille, 105, 113
OnePlus, 224
OnlyFans, 274–276, 286
@oscarprgirl (Twitter account), 110, 112–113
Ostrenga, Kirsten (Kiki Kannibal), 32–33, 35–37
Otter Media, 163

PageSixSixSix, 16
paid partnerships, 217–218, 225
Pakman, David, 234
Palermo, Olivia, 4–6, 219
Pappas, V., 92
parasocial bonds, 35
Parcell, Rachel, 223

Parker, Trey, 116
Parnell, Chris, 60
Paternot, Stephan, 30
Patreon, 240, 276, 286
Paul, Aaron, 239
Paul, Jake, 177, 210–211, 237, 240–242, 249, 250, 284
Paul, Logan, 151, 167, 177, 192, 196, 197, 199, 210, 211, 237–242, 249
Payne, Candace (Chewbacca Mom), 194
PayPal, 59
PCA Skin, 215
Pence, Mike, 277
Penchansky, Raina, 223
Penna, Joe (MysteryGuitarMan), 169
Pepper, Sam, 211
Peretti, Jonah, 48
Periscope, 158, 161
persona, 109–110
personal branding, 42–43, 45
Petrou, Thomas, 268
PewDiePie, *see* Kjellberg, Felix
Phan, Michelle, 90
Piker, Hasan, 277
Pinterest, 138
Pioneer Woman, 24
Piques, Jon Paul, 186, 193, 196, 197
Pizzagate conspiracy, 278
Platco, Michael, 199, 201
Poarch, Bella, 266–267
point-of-view videos, 274
politics and political content, 12–16, 165, 231–232, 276–278

Pons, Lele, 177, 187, 196, 197, 249, 250
Popular page, 131, 178–180, 185
Portnoy, Dave, 270
Power, DeStorm, 169
power users, 130, 132, 186
prank videos, 191–192, 210, 211, 229–231, 243, 245–246, 249, 250, 285
pre-edited content, on Vine, 178, 180
Price, Cece, 271–272
Procter & Gamble (P&G), 23
Pseudo.com, 157
Pugh, Sydney, 224
Pulse program, 285
Punk'd (TV series), 43
Purina, 95

QAnon, 278
Quinn, Zoë, 164–165

radicalization, 164–165
Rahma, Kareem, 278
Rankzoo, 177, 183
reality television, 36, 42–46, 51, 52, 198, 271–272
Real World, The (TV series), 43–44
Red Bull, 133
Reddit, 53, 94, 120, 164, 231, 247
Reeves, Anthony, 269
referral codes, 222
Rei, Olga, 5
relatable YouTube, 258–259
Relles, Ben, 78
remote learning, 274

Renegade dance, 279–281
Rensing, Emil, 77
reposting, 48, 107, 115–117, 178, 184, 197, 198
@therepublicanhypehouse (TikTok account), 277
reputation, digital, 290
Revlon, 90
Revver, 73–74
RewardStyle, 137–138, 221, 284
Reynolds, Carter, 171
RiceGum, 210, 211
Richards, Josh, 269, 270, 285
Richie, Nicole, 44
Riedel, Josh, 130, 132
Rivera, Brent, 175–176
ROFLCon, 49
Rojas, Christopher, 169
Rose, Jessica Lee, 63–65
Rushfield, Richard, 65
Rutherford, Chrissy, 281, 282

Sadowski, Josh, 270
Sala, André, 181
Salomon, Rick, 44–45
Samberg, Andy, 60
Sanyo, 80
Sartorius, Jacob, 206
Saturday Night Live (TV series), 60–61
Scannell, Herb, 77
Scare PewDiePie (web series), 229
Scene Queens, 32–33, 35–36, 250
Schaffer, Akiva, 60

schedule, creators', 243, 245–247, 250–251, 258, 268
Schindler, Philipp, 233
Schlossman, Lawrence, 118
Schmidt, Eric, 67, 71, 72, 74
Schnipper, Lauren, 195
scripted content, on YouTube, 63–64
security, for meet-and-greets, 172–173
Segal, Barrie, 280
Seibel, Michael, 152, 157
Seibert, Fred, 77, 92
selfies, 140–141
self-parody, 252–255
self-promotion, 46–48, 135, 140, 241–242, 275
Sequoia Capital, 60
"Sex on the Hilltop" (column), 41
sex-tape scandal, Paris Hilton, 44–45
Shade Room, The, 271, 281
Shahidi, John, 176–177
Shahidi, Sam, 176
shareable content, 279, 280, 290–291
ShayCarl, 79
Shear, Emmett, 157
Sheffer, Sam, 268
Shey, Tim, 48, 77, 78, 92, 93
Shorts, YouTube, 285
Shot at Love, A (TV series), 36
Shots app, 176
Sideman, Adi, 157–158
Simmons, Jed, 77
Simon, David, 17
Simple Life, The (TV series), 42, 44–45
Simpson, Jessica, 43

SixDegrees.com, 30
1600 Vine crew, 175–180, 184–188, 191–192, 197, 198
#sleepsquad, 158
slut-shaming, 51–53
Smalls, Joan, 141
Smiles, Jessi, 167
Smith, Brodie, 167
Smith, Paul, 212
Smith, Phil, 233
Smosh, 75
SMS text messaging, 103
Snapchat, 152–154, 166, 168, 188, 193, 194, 197–202, 216, 285
snark, 247
Snoop Dogg, 128–129
Socialcam, 152, 180
Socialite Rank (blog), 2–7, 219
social media creator content houses, *see* collab houses
social media editors, 178–179
social media platforms (generally)
    adult content on, 275
    connections with friends on, 29
    creator relationships with, 29, 70, 82–83, 87–89, 162–164, 179–188, 197–200, 208–209, 289–290
    influence metrics on, 93–94
    Instagram links to, 128
    redirecting of, 85
    rise of, 5–7
    transferring audience to other, 191–193
    treatment of creators by, 29, 289–290
    users' influence on, 1–2, 287, 289
    weaponization of, 164–165
    see also specific platforms
Social Media Week event, 49
social networks, 29–40, 48–49, 105
    see also specific platforms
Something Navy, 220–222
Sony, 51
South by Southwest (SXSW), 95, 105, 117
Southern Methodist University, 136
Spacestation, 202
spam, 139–140
Spencer, Karyn, 182–188, 191
Spiegel, Evan, 153, 200
sponsored content deals, 25–26, 35, 66, 119, 133–136, 139–140, 165–168, 197, 199, 206, 211, 215–222, 224–225, 246–247, 253–254, 276, 281
Spotlight, 285
Star, Jeffree, 248
Starbucks, 133
StarCraft (video game), 158
Station (channel), 79–81, 87, 163, 175
status, 39, 93–94, 102
Stevens, Michael (Vsauce), 78
Stewart, Lance, 193
StickyDrama (blog), 36–37
Stokely, Tim and Thomas, 275
Stone, Biz, 102, 104, 107
Stone, Christopher, 36

Stone, Lisa, 24
Stone, Matt, 116
Stories feature, 153, 168, 198, 199, 201, 220, 222, 254, 257
Strompolos, George, 61, 67, 68, 74, 75, 80–82, 87–91, 96–97, 112, 164, 209
Studio71, 91
Suave, 23
subscription revenue, 240, 275–276
Substack, 286
suggested user list, Instagram, 135
suicide, 237–240
Sway boys, 269–271, 284, 285
Swift, Taylor, 122
Systrom, Kevin, 127–135, 139, 140, 152

Taccone, Jorma, 60
Taco Bell, 213
TalentX, 269
Talking Points Memo, 12–15
Target, 25
Tay Zonday, *see* Bahner, Adam
Team 10 house, 211, 212
Tequila, Tila, 31, 35, 36
terrorist groups, 232
Thompson, Clive, 103–104
Thurmond, Strom, 14
TikTok, 40, 53, 92, 145, 153, 208, 247, 263–274, 276–285, 290
 see also Musical.ly
TikTokalypse, 270–272
@TikTokRoom (Instagram account), 271, 280

TMI Weekly (web series), 78
T-Mobile, 51
Today show, 151
Toff, Jason, 180–181, 188
Topshop, 219
touring, by creators, 168–173
transparency, 217–218
Treasure & Bond, 220
TRESemmé, 220
Trident, 167
Triller, 285
Trump, Donald, 165, 197, 231–232, 276–278
Truth in Advertising, 216
Tsuki (merchandise), 242
Tumblr, 48–50, 53, 78, 93, 101, 115–125, 146
Turner, Suzie, 222
Twaimz, Issa, 211
Twitch, 157–159, 161, 207, 276–277
Twitter, 38, 39, 48, 93, 95, 101–114, 120, 122, 128–129, 140–141, 146–147, 166–168, 182, 183, 186–189, 231–232, 285

Uhovski, Valentine, 5
Unilever, 51
United Talent Agency, 36, 223
Urlesque, 49–50
user-generated media, 11–12, 157–158

Vaccarino, Chris, 241, 242
Valiando Rojas, Meridith, 169, 172
validation, 224–225, 291

372 // INDEX //

Valid Crib, 281, 282
Van Brocklin, Cara Loren, 215
vanity URLs, 179
Vaynerchuk, Gary (Gary Vee), 166
venture capital, 60, 89, 91, 280, 283–284
Venz Box, Amber, 136–139
Verizon, 25, 233
Viacom, 172
VidCon, 85–87, 180–182, 195, 206, 208
video apps, 151–159, 193, 235–237
    see also specific platforms
views culture, 249
Vimeo, 53
Vine, 145–155, 162, 163, 165–171, 175–189, 191–202, 205–213, 265, 267
Viner of the Year party, 183–185
Violet, Alissa, 211
viral content, 47, 61, 68–73, 80, 95, 107–108, 111, 115, 116, 119, 120, 122, 131, 167, 170, 193–195, 242, 266, 274, 281
Vlogbrothers, *see* Green, Hank; Green, John
vlogging, 62–63, 153, 209–211
Vlog Squad, 285
Vogt, Kyle, 157
VYou, 157

Walk, Hunter, 87
wall scouting, 212
Walmart, 286

Walt Disney Company, 163, 164, 229
Walters, Barbara, 43
Warner Music Group, 207
Warren, Alex, 268
watch time (metric), 209, 210
Wayfair, 278
webcams, 62–63
web logs, *see* blogs
Webutante Ball, 49
Weichel, Sarah, 121, 123
WhataDayDerek, 79
Wheelhouse, 271
Whitney, Shea, 222
William Morris (agency), 163
Williams, Evan "Ev," 102, 104–108
Winfrey, Oprah, 51, 108
Wish, 270
Wojcicki, Susan, 61, 66–67, 237
Wolfe, Tom, 42
women, 22–24, 26–27, 36–37, 51–53, 164–165, 186, 223, 250, 284
    see also mommy bloggers
Wonderful Pistachios, 72–73
Woolf, Rebecca, 19–21, 25
Worldstar, 281

Yahoo!, 125
Yang, Luyu, 153
Yglesias, Matthew, 175
YouNow, 157–158, 161–162, 169, 172, 206
YouthStream Media Networks, 30
YouTube, 59–83, 166
    Adpocalypse at, 231–235

// INDEX // 373

advertising on, 78, 80–82, 89–90, 97, 182, 186, 231–233
and aggregators, 120
backlash against prank videos on, 229–231, 285
Justin Bieber on, 130
collab channels on, 76–80, 267
creator relationships with, 87–88, 208–209
and DigiTour, 169
disclosure of sponsorships on, 217
Elsagate on, 235–237
entertainment industry and, 162–163
fan–creator relationships at, 85–87
founding of, 59–60
front page of, 68–69, 81
and Google, 61–62, 66–67, 207
gossip/drama channels on, 247–250
growth of, 60–61, 66, 76
harassment of creators on, 250–252
haul videos on, 222
lo-fi content on, 255–259
Lonelygirl15 series on, 62–66
MCNs on, 166
merchandise deals for creators, 240–243
monetization on, 39, 71–76, 197
multichannel networks at, 82–83, 89–93
Musical.ly and, 206–207, 265
Partner Grants Program at, 92

Logan Paul controversy at, 237–240
redirecting of, 85
resharing on, 178
schedule for creators on, 244–247, 250–251
Snapchat and, 202
subscriber metrics on, 88–89
and TikTok, 270, 271
value of creators at, 96–97
VidCon presence of, 181
Vine and, 146, 148, 188, 192, 193, 208–211
viral content on, 53, 68–73, 80
YouTube Live event, 76–77
YouTube Next, 92
YouTube Partner Grants Program, 87–88, 92, 208
YouTube Partner Program, 67, 72, 74–77, 82, 87–88, 96–97, 209, 210, 289
YouTube Spaces, 208
Yung Klout Gang, 93–94
Yusupov, Rus, 146, 151, 183

Zamora, Gabriel, 248
Zappin, Danny, 77, 79–80, 82, 90
Zhu, Alex, 153
Zimmerman, Neetzan, 122
Zoella, 169
Zombies (TV series), 264
Zoom feature, 257
Zuckerberg, Mark, 33, 34, 37, 39, 113, 193, 194